HART LICHTENWALNER'S COPY
PASS ON TO FRANK BREWSTER
AND RON WILT.
WHEN FINISHED, RETURN TO HART

FRANK POTOCHNIK

UNLIKELY VICTORY:

HOW GENERAL ELECTRIC SUCCEEDED

IN THE

CHEMICAL INDUSTRY

BY JEROME T. COE

AMERICAN INSTITUTE OF
CHEMICAL ENGINEERS
www.aiche.org

Published by the American Institute of Chemical Engineers
3 Park Avenue, New York, NY 10016-5991

In Association With
The Chemical Heritage Foundation
315 Chestnut Street
Philadelphia, PA 19106-2702

© 2000 by the American Institute of Chemical Engineers
3 Park Avenue, New York, New York 10016-5991
www.aiche.org

Library of Congress Catalog Card Number 00-036229
ISBN: 0-8169-0819-2

Coordinating Editor: Beth Shery Sisk
Cover Design and Typesetting: Terry A. Baulch
Copy Editing: Trumbull Rogers

Unlikely victory: how General Electric succeeded in the chemical industry
by Jerome T. Coe.
    p.   cm

    Includes bibliographical references and index.
    1. General Electric Company. 2. Chemical Industry - - United
States. 3. Conglomerate corporations - - United States.
4. International business enterprises - - United States. I. Title.

HD9651.9.G43.C63   2000
338.7'66'00973--dc21   00-036229

Printed in the United States of America.

Table of Contents

Chapter   Page

Preface v

Acknowledgments ix

1. What's General Electric Doing in the Chemical Business? 1

2. Early Years of GE Chemistry: 1900-1948 9
    Electrical Insulation; Silicones; GE Forms a Chemical Division

3. GE Silicones: 1940–1964 27
    From Shaky Start to Successful Business

4. Loctite 45
    An Invention that Got Away

5. Synthetic Diamond 49
    GE Break-Through Caps Two Centuries of Research

6. Lexan Polycarbonate: 1953-1968 69
    The "Unbreakable" Thermoplastic

7. Noryl Thermoplastic: 1956-1968 83
    Victory Snatched from Jaws of Defeat

8. GE Engineering Plastics: 1968-1987. 91
    Headlong Growth to World Leadership

9. Growth by Means of a Major Acquisition: 1988-1991 113
    ABS Plastics Up for Bid; A New Polycarbonate Process

10. Laminates and Insulating Materials. 123
    GE Core-businesses Decline in Importance

11. GE Silicones: 1965-1998 129
    Sealants Leadership; World Participation

12. GE Engineering Plastics: 1992-1998. 139
    After Recession, Growth Resumes

13. People Make the Difference 159
    *Four Scientists: Eugene G. Rochow, H. Tracy Hall and the
    GE Diamond Research Team, Daniel W. Fox, Allan S. Hay
    Five Managers: Abraham L. Marshall, Charles E. Reed,
    John F. Welch, Jr., Glen H. Hiner, Gary L. Rogers*

14. Summation 178
    How Big an Achievement? How Attained? Nine Strategies

Glossary 195
A. *Thermoplastic Polymers, Compounds, and Blends* 195
    B. *Trade-names, Companies, and Chemical Terms.* 196
    C. *GE Organization Notes* 199

Chapter References. 201

Names Index 211

Subject Index 215

Preface

The path by which General Electric (GE) became a major factor in the chemical industry is a fascinating story for those of us who worked in that part of the company and who have followed its progress in later years. GE's chemical industry products—engineering plastics, silicones, and synthetic diamond make up one of the major growth areas for the company in the post–World War II era, the notable others being financial services and aircraft jet engines. These three nonelectrical segments now account for well over half the company's revenues and earnings. The chemical products are not only an important part of GE, but they place it in the top ranks of U.S. chemical industry producers.

I will review the major chemistry discoveries, technology developments, marketing programs, and the management initiatives, or strategies, which together created the large and successful endeavor now called GE Plastics. I will also chronicle some individuals who played key roles in these teamwork efforts.

My GE career began in 1942 in the Research Laboratory, where I joined the silicone project as a chemical engineer after graduating from M.I.T. After service in the navy, I rejoined GE in 1946. The company was then building a silicone plant at Waterford, N.Y. After some years managing silicone process development and manufacturing engineering, I headed sales development, then field sales, and finally marketing. I was general manager of GE Silicones from 1959 to 1964, after which I led GE's computer services business (called "time-sharing"), and then I headed a division that included diamond manufacturing. Strategic planning and corporate staff assignments completed my 40-year GE career.

My point of view in this history, which describes GE's chemical industry experience during more than half a century, is that of general management: how are the people, technology, and investment assets of a business employed to serve customers profitably in a market, in competition with other organizations equally well staffed and financed? What discoveries were made? What initiatives were chosen by the managers? How well did these choices work out? What lessons did GE managers learn and how were these applied in later programs? The principal measures of the success of such efforts are sales, sales growth, share of the market, and the rate of income return on sales and on assets. While this is a General Electric story—GE disappointments as well as accomplishments are described—I have tried to present competitive perspective realistically.

This work is not a company-authorized history, although top executives encouraged me and have cooperated with interviews and assistance. Judgments on content and emphasis are my responsibility, as are inadvertent errors.

I hope this book will serve three groups of readers:

1. Chemists, chemical engineers, industry executives (and their former professors), who are interested in how GE's chemical discoveries and technology developments were profitably matched to market opportunities; in short, "How'd they do that?"

2. Students and teachers of business management practice attracted by the management lessons GE learned along the way.

3. General Electric employees, present and retired, who want to know, or remember, how this part of the company evolved and became successful.

GE now refers to its major businesses descriptively, for example, GE Plastics, GE Aircraft Engines, GE Lighting; and the leaders of the large units are now titled President and Chief Executive Officer. *GE Plastics* includes the present chemical products of this history. GE Plastics' overall financial results have been visible since 1984 when the company's annual report first presented data for a *Materials* business segment. This segment's title was changed to *Plastics* with the 1998 annual report. This segment's data are essentially synonymous with *GE Plastics,* as well as with my descriptor, GE's chemical industry products. GE does not regularly release financial information on product lines in greater detail than the annual report breakdown, but enough news articles, papers, books, oral histories, and personal recollections are available from which to construct a coherent history of GE Plastics with some financial perspective. Whereas the annual reports refer to an *engineered* plastics product line, I have used the more common industry term, *engineering* plastics.

Not every chemical or process industry participation by GE is included here. For example, outside this book's scope GE Lighting makes phosphors for fluorescent lamps, processes ore for making tungsten and tungsten wire, and makes the special ceramic used in high-intensity discharge lamps. In addition to melting quartz to form special lamp envelopes, GE Lighting also sells quartz crucibles that are widely used in hyper-pure silicon zone-refining furnaces.

In the 1930's both the GE Schenectady and Pittsfield Works formulated, formed, and baked large ceramic insulators needed for electric power generation, distribution equipment, and electric utility transmission lines. These insulator plants were later closed, but GE acquired majority ownership of the Locke Insulator Co. of Baltimore, Maryland. This company was finally sold to Japanese interests.

General Electric succeeded Du Pont in 1946 as contractor to the Atomic Energy Commission for the Hanford, Washington, nuclear facilities, and resigned that role in 1964. At one time GE designed and built for the AEC a spent nuclear fuel reprocessing plant that was subsequently shut down. GE's commercial nuclear energy business fabricates nuclear fuel reloads for existing power plants and for the occasional new facility.

A GE department once made space reentry vehicles, and developed pyrolytic graphite carbon polymers to protect the vehicle entering the atmosphere.

The Loctite chapter is included, not just because a GE research discovery and an alumnus played a part, but also because the Loctite Corporation's chemical innovation success demonstrates management lessons similar to those from GE's product experiences.

I have not dwelt on all the (sometimes bewildering) changes in GE organization and its nomenclature, but have mentioned those most relevant to this chemical business. The early Research Laboratory became Corporate Research and Development Center, or C. R.&D. I have used these terms interchangeably. Works Labs became Materials and Processes Labs, but I have stayed with the earlier term. Chemical *Division* is used here rather than its earliest designation as *Department*. Chemical Development *Operation* is cited rather than the earlier Department. (Also see Appendix C on GE organization.)

No attempt has been made to trace all personnel changes in various positions, but the path of some individuals is presented so their GE background or their progress to other parts of the company can be understood. Contributions of many people will be cited, and brief biographies of key scientists and managers are presented.

Business strategy is the sum of decisions made by the business managers, the term usually meaning decisions with long-term impact. A successful strategy, competently carried out, establishes a competitive advantage of some duration. Business success flows not only from the strategy chosen, but also from organizational competence to carry out the plan. So both the game plan *and* the execution are critical. GE's effectiveness in managing chemical businesses grew greatly with experience, as this book demonstrates. Scientific discovery, technology development, creative marketing, and management strategies all played a role in the success. I have emphasized those discoveries, events, and strategies that were most important to the present outcome.

Jerome T. Coe
Greenwich, Connecticut
May 15, 1999

Acknowledgments

I am deeply indebted to many individuals for their contributions to this book through recollections, referrals, comments on drafts, writings, and formal publications.

George Wise, formerly of GE's Corporate Research and Development Center, has been especially helpful with editorial suggestions, references, and historical knowledge of both the R&D and commercial aspects of GE's chemical business. He is the author of *Willis R. Whitney, General Electric, and the Origins of U.S. Industrial Research* (Columbia University Press, 1985). He also provided the complex history of diamond making, including the early GE discoveries.

Charles Reed, who retired as Senior Vice-President–Technology of General Electric in 1979, gave me much encouragement and shared extensive recollections from his 37 years with GE, most of them associated with the silicones, diamond, and engineering plastics businesses.

Jack Welch, GE's Chief Executive Officer, encouraged me to write the book, as did Gary Rogers, CEO for GE Plastics, and each granted interviews.

Chemical Heritage Foundation Oral Histories from Daniel W. Fox, Allan S. Hay, Charles E. Reed, Eugene G. Rochow, J. Franklin Hyde, and Earl L. Warrick, were an important resource, as were Liebhafsky, *Silicones Under the Monogram*, (John Wiley, 1978) and Morone, "New Business Development and General Managerial Decision Making," *Research on Technological*

Innovation, Management (JAI Press, 1993). Leonard Fine, author of *Chemistry for Nonchemists, A Practical Guide to the Science and Technology of GE Plastics* (unpublished) provided several important personal recollections. These and many other relevant writings, both published and unpublished, are noted in the Chapter References section.

The book's content editor, Otto T. Benfey, made significant contributions that improved the clarity of presentation, arrangement of subject matter, and technical details throughout the whole document.

I'm indebted to Stephen Smith and Gail Nalven of AIChE Publications for early encouragement and later commitment to publish this work of a first-time author. Many thanks also to the AIChE team of Beth Shery Sisk, Trumbull Rogers, and Terry Baulch for bringing the manuscript to publication.

It has been gratifying to renew so many acquaintances with friends and former business associates. I was able to make some contacts in person, but many were by lengthy telephone conversations. Individuals who made special inputs are noted with the references for each chapter, and I am grateful to all the following who provided input, insights, and help:

Pier Abetti, Raymond Andresen, Werner Bachli, Albert Baldock, John Batty, James Beach, Swede Berntson, Harold Bovenkerk, Janie Castles Boyd, Donald Brown, Richard Brown, Francis Bundy, Joseph Burke, Robert Canning, Joseph Caprino, William Christopher, Joseph Cohen, William Cordier, James Costello, Robert Daily, Dean Daniels, Paul Dawson, Wayne Delker, Walter Dugan, Leonard Fine, Eva Fowler, Robert Friedman, John Garrison, Thomas Gensler, Albert Gilbert, Stokes Gillespie, Wyman Goss, Philip Gross, David Guilbault, Reuben Gutoff, David Hall, George Hartley, Alex Hasson, Robert Hatch, Allan Hay, Robert Hess, Glen Hiner, Jean Heuschen, Robert Imsande, Phil Jeske, Reginald Jones, Joan Kadlec, Louis Kapernaros, Louise Koval, John Kennedy, Thad Leister, Hart Lichtenwalner, Buzz Lippincott, John Loritsch, Jeffrey Lott, Joseph Lott, Marie Lynn, Paul McBride, George McCullough, E. Charles McClenachan, Mary Ann Merlini, Donald E. Miller, Donald H. Miller, Richard Moeller, Leroy Moody, Joseph Morone, Robert Muir, Kevin Murray, Gail Nalven, Diana Nichols, Theodore Ohart, Frank O'Keefe, John Opie, Jack Peiffer, Marianne Poulliot, James Pyle, Charles Reed, Eugene Rochow, Gary Rogers, Mrs. Popkin Shenian, William Sessions, Tillar Shugg, Eva Smith, Stephen Smith, Meg Spencer, John Stauffer, Herbert Strong, William Viscusi, William Vivian, Beth Vodapivc, Jack Welch, Jeff Welde, Robert Westervelt, Steven Williams, Joseph Wirth, George Wise and Ronald Wishart.

I am very grateful to my dear wife, Peggie, who encouraged me during several years of effort, and who also launched me toward computer competence, without which the task would have been impossible.

CHAPTER 1

What's General Electric Doing in the Chemical Business?

H ow did General Electric, an electrical manufacturing company, become one of the largest and most successful U.S. chemical companies?

Successful long-term diversification is rare among giant U.S. companies. Many tried: Exxon for a while sold electric motors and office computers, but is now again an oil and petrochemicals company. General Motors tried refrigerators, Ford tried electronics, and Chrysler made tanks and corporate jets; they are now more narrowly focused car companies. AT&T, after long ago being a competitor of GE in electrical machinery and later making a major try in computers, is again a telephone company. It has recently spun off its equipment design and manufacturing business, and is acquiring cable TV companies.

The conventional wisdom has usually argued against wide corporate diversification. Andrew Carnegie in the 1880s said, "Put all your eggs in one basket and watch that basket." Tom Peters and Robert H. Waterman in their 1980s bestseller *In Search of Excellence,* said "Stick to the Knitting."[1] For a time in the 1960s, conglomerates became fashionable, but the subsequent difficulties of such leaders as Textron, RCA, and ITT only underlined the earlier wisdom.

GE has spent over a century ignoring that conventional wisdom. Its businesses now include financial services, currently the largest earnings contributor, aircraft jet engines, television broadcasting, information services, plastics, medical equipment, as well as light bulbs, turbines, transformers, switchgear, electric motors, and appliances. It is now classified by *Business Week* as a Conglomerate. Compare it with the others on this magazine's (1996) conglomerate list.[2] GE's sales are more than five times as great as the next largest conglomerate, Allied Signal, and its earnings are more than twice as large as all the other eleven conglomerates on the list put together.

So GE's diversity is both successful and surprising. How was it accomplished? GE did not always achieve diversity from strategy, but sometimes from necessity. In chemistry, for example, it became good at making polymers in order to keep electricity in wires. Those chemical interests eventually stimulated research in the field of engineering plastics, the high-performance end of moldable plastics. Then, as much to its own surprise as anyone else's, GE found itself in a position to grow a worldwide business, so technically complex and so capital-intensive that competitive entry was difficult.

It was not the outcome of a conscious long-range plan. It was not directed from corporate headquarters by a bureaucracy. It was guided for some decades by a small number of individuals acting as a small business, tucked away in a corner of the corporation. Much of what they did was counter to the then prevailing GE culture. Today it has become the corporate culture. In part, this is because one of those pioneers in the GE hinterlands, Jack Welch, is now GE's Chairman and Chief Executive Officer. Not many remember that John F. Welch, Jr., got his start in the GE plastics business, which he joined after receiving a Ph.D. in chemical engineering from the University of Illinois.

Although the magnitude of GE's chemical participation is not widely known, competitors in the industry niche called "engineering plastics" are well aware that GE is the world leader in these thermoplastics. Within GE, recognition of the size, scope, and contribution of this business was scarcely noticed until fairly recently. But beginning with 1984, GE annual reports[3] show segment financial data for "Materials," now retitled "Plastics," which comprises: *High-performance engineered plastics used in applications such as automobiles and housings for computers and other business equipment; ABS resins; silicones; superabrasive industrial diamonds; and laminates. Sold worldwide to a diverse customer base consisting mainly of manufacturers.*

GE's 1998 sales from these chemical-type products were $6.6 billion, operating profit was $1.58 billion, and assets employed were $9.8 billion. The recent Plastics sales are comparable to GE's aircraft jet engines, major appliances, or power-generation businesses. Plastics' profit contribution has been comparable to aircraft engines and is greater than power generation, appliances, NBC broadcasting, or lighting products. Assets employed in these chemical-type businesses (year-end 1998) are greater than any other GE manufacturing segment, showing both that the chemical businesses are capital-intensive and that GE has been willing to invest heavily in this arena.

In 1948, three years after a GE Chemical Division was first formed, its contribution to company *external* sales was around 1.5% and to profits was nothing. So over a 50-year period, GE's chemical business grew from negligible contribution to a 7% sales and a 12% profit share within the corporation. In the same period, General Electric's total revenues grew 61-fold to $100.5 billion, earnings grew 66-fold, and its stock market valuation reached number one among U.S. corporations (currently No. 2 behind Microsoft).

The organization now responsible for these products is *GE Plastics*, which includes not only the engineering plastics and ABS resins but also *Polymerland*, a distributor of many plastics, a petrochemical styrene producer, and a phosphite plastics additives product line. *GE Silicones, GE Superabrasives*, and *GE Electromaterials* are separately managed businesses, also within the *GE Plastics* structure.[4]

Engineering plastics and the related products and services are by far the largest and most important of the GE Plastics products. These thermoplastics are specially crafted by chemical polymer engineering to meet one or several high-performance specifications, such as strength; heat and cold resistance; impact strength; dimensional stability; oil, solvent, or moisture resistance; durability in sunlight; and many other properties needed for different applications in the manufactured-products world. Engineering plastics prices are higher than those of the high-volume commodity plastics and are produced in much smaller tonnages. The automotive, appliance, computer, telecommunications, electronics, and construction industries are major end users of engineering plastics parts.

GE is the overall global leader in engineering plastics by a wide margin, almost twice as large as the nearest producer. The major world competitors in this niche include Du Pont, Celanese, Bayer, BASF, Dow Chemical, and Mitsubishi Gas Chemical.

The most important GE plastics product lines are:

LEXAN polycarbonate resins (PC), plus glazing and film;

NORYL modified polyphenylene oxide (PPO) resin alloys;

VALOX polybutylene terephthalate resins (PBT);

XENOY PC/PBT resin alloys;

CYCOLAC acrylonitrile/butadiene/styrene (ABS) resins;

CYCOLOY PC/ABS alloys;

ULTEM polyetherimide (PEI) resins.

World headquarters for GE Plastics is Pittsfield, Massachusetts. Major polymer manufacturing plus compounding facilities are located throughout the world: in the United States (8 locations), The Netherlands, Spain, Germany, and Japan (4 locations). Additional plastics compounding is conducted in Canada, Mexico, Brazil, Argentina, Scotland, France, Italy, South Korea, Singapore, China, and India. In addition to the Pittsfield and Netherlands headquarters, twelve commercial development technical centers are located in the major industrial countries of the world.

GE Silicones is another important segment of GE Plastics. The product lines comprise dimethylsiloxane fluids of many viscosities and various volatilities, fluid-water emulsions; silicone rubber gums and compounds; sealants and caulking materials, plus other liquid-rubber products from a family of room-temperature vulcanizing (RTV) silicone compounds; silicone resins in solvent solution; and many other silicone specialties. GE Silicones headquarters and main plant are at Waterford, New York. Rubber compounding capacity is also located in The Netherlands. GE Toshiba Silicones (Tosil) is a joint venture in Japan, as is GE Bayer Silicones in Europe.
The world volume leader in silicones by a wide margin is Dow Corning, with GE (including the Tosil and Bayer joint ventures) in second place. Other major world producers include Shin Etsu (Japan), Wacker-Chemie (Germany), Rhodia (France), and Witco.

Superabrasives include a range of small synthetic-diamond crystals, the finer mesh sold to make diamond grinding wheels, and the larger to manufacture diamond saws. Diamond polycrystalline compacts, or blanks, made in various shapes from cemented diamond crystals, are used to make wire-drawing dies, special cutting tools, and oil exploration bits. Borazon, a cubic boron nitride, forms special grinding wheels, and Borazon compacts machine special steels. The GE Superabrasives headquarters and main plant are at Worthington, Ohio, and a second major plant is in Ireland. GE is considered the world leader in synthetic-diamond products, but De Beers Consolidated Mines, Ltd., is a major competitor worldwide in these superabrasives.

The *laminates* products offered by *GE Electromaterials*, are layers of glass fiber bonded by various resins and formed under pressure and heat into tough sheet materials having excellent electrical insulation value. The laminates are usually copper clad to form baseboards for electronic printed circuits. The sales contribution of this product line is now small in the GE Plastics total. GE Electromaterials headquarters and plant are at Coshocton, Ohio.

When the 1984 GE Annual Report broke out the chemical business data for the first time, many within GE were surprised by its size and earnings contribution. The breakout also made it possible to assess GE's chemical activity in relation to others in the chemical industry. *Chemical and Engineering News* has since 1968 published a special ranking of the leading U.S. chemical producers (originally 50, then 100, now the top 75).[5] The data are derived from annual reports of companies with chemical-type businesses or product lines. The *Chemical and Engineering News* presentation subtracts from corporate totals such segments as petroleum industry participation, metals production, and consumer product sales, but it includes chemicals, fibers, and plastics. For Du Pont, its chemicals, fibers, and polymers segments have been included, but not its petroleum, pharmaceutical, or diversified businesses. Dow Chemical data have included all segments except its consumer specialties. Monsanto's data in this analysis (prior to the Solutia spin-off) included its Agricultural and Chemical Groups, but not Searle or NutraSweet. The magazine cites the "plastics" data above for General Electric.

The *Chemical and Engineering News* listings of U.S. chemical producers in 1968 and in 1998 by chemical sales show the following:[5]

Rank	1968	1998
1.	Du Pont	Du Pont
2.	Monsanto	Dow Chemical
3.	Union Carbide	Exxon
4.	Dow Chemical	**General Electric***
5.	W. R. Grace	Union Carbide
6.	Exxon	**Huntsman***
7.	Celanese	**ICI America***
8.	Allied Chemical	**Praxair***
9.	Hercules	**BASF***
10.	Food Machinery	**Eastman Chemical***

While this ranking in recent years has been affected by spin-offs, mergers, and acquisitions, General Electric is currently (1998) the highest ranked of the six asterisked new entries to the top-ten chemical sales list.

In 1968 GE's chemical-type product sales were around $300 million, which, had they been publicly visible at the time, would have made a *Chemical and Engineering News* ranking of No. 29. By 1984, GE had become a more significant player in the chemical industry: in that year's analysis GE ranked No. 15 in sales, No. 6 in operating profit, No. 9 in assets, and was shown to be a leader in both ratio of profit to sales and profit to assets.

In 1993, surprisingly, GE was first in profit:[5]

1993 ($ mills)	Chemical Sales	Chemical Operating Profit	Chemical Assets
Du Pont	$15,603	$750	$13,957
Dow Chemical	12,524	773	12,530
Exxon Chemical	10,024	638	8,478
Hoechst Celanese	6,347	561	NA
Monsanto	5,651	731	5,312
General Electric	**5,042**	**834**	**8,181**
Union Carbide	4,640	354	4,419

The 1993 comparisons are not typical for the major competitors, as a recession that year depressed profits severely for many. The much larger chemical sales of Du Pont, Dow Chemical, and Exxon Chemical will ordinarily prevail in the profit column also. But recent GE Plastics sales and operating profit consistently rank high among U.S. chemical producers:

GE	Chemical Sales Rank	Chemical Operating Profit Rank
1998	4	4
1997	4	4
1996	6	4
1995	6	5
1994	6	4
1993	6	1
1992	7	2

Special recognition of these industrial achievements came with a 1991 National Medal of Technology award to Charles E. Reed, a retired senior vice-president of General Electric, whose 32-year career with GE was largely devoted to the creation and growth of the company's chemical business. This medal is presented annually by the nation's president to individuals and companies for their outstanding contributions to improving the well-being of the United States, either through development or commercialization of technology or for their contributions to the establishment of a technically trained work force.

Reed's medal citation reads:[6]

For his management risk-taking in continuous innovation leading the General Electric company to world class production of advanced engineering materials.

Starting from a negligible industry position in 1945, the magnitude of GE's success was unexpected both inside and outside General Electric, because GE's chemical business developed slowly from a faltering start. But as competitive competence increased within the organization, GE marketed innovations in *silicones, synthetic diamond,* and then *three engineering plastics,* becoming leader in the latter niche of the plastics industry and reaching the top ranks of U.S. chemical producers. The remarkable outcome to this point stems from technical innovations, successful leadership, and manage-

7

Fig. 1-1. Charles E. Reed (right), a 1991 National Medal of Technology recipient from President George Bush (left).

ment strategies that were implemented by the organization more and more effectively as time went on. In the 50-year effort a large number of scientists, engineers, salesmen, and managers all made essential contributions to the success of these high-technology businesses.

By entering the chemical business and investing heavily in it over the years, GE clearly did not "stick to the knitting," at least as knitting would be defined by the company's product scope of 1945. For a long time the chemical effort was not a big corporate success, even skirting failure in the early years. But, in the long run, General Electric's rise to the top ranks of the U.S. and world chemical industry is a remarkable and unexpected achievement. How this unlikely victory was accomplished is the story of this book.

CHAPTER 2

Early Years of GE Chemistry: 1900–1948

Electrical Insulation; Silicones; GE Forms a Chemical Division

G eneral Electric Company emerged as a major corporation because Thomas Edison did more than just invent a better light bulb. He embedded it in a system for lighting and powering the world. The system created the electric utility industry. The supplying of equipment for the system created GE, founded as a merger of Edison's company and some rivals in 1892.

GE became involved with chemistry because the wires, connectors, and equipment that carried and used electricity had to be insulated: that is, covered with something that kept the electricity in. Edison's workers used various kinds of natural materials or their derivatives to do this job: wood oil, shellac, mica, paper, cloth, and rubber. GE continually searched for new solids, liquids, and rubber materials that would block the leakage of electric current. The effort was carried on in the corporate Research Laboratory and in labs attached to GE plants (called Works) in Schenectady, New York; Pittsfield and Lynn, Massachusetts; and Ft. Wayne, Indiana.

In 1906, an independent inventor, Leo Baekeland, invented a thermosetting plastic with unusually good strength, rigidity, and electrical insulating characteristics. This material, later trade named Bakelite, has often been called the world's first truly synthetic plastic. Other plastics had been made by re-forming natural materials, but this one started from simpler chemicals, phenol and formaldehyde. Both as a liquid resin and as a molding com-

9

pound, Bakelite was an improvement over its natural predecessors.

In 1909, Baekeland brought his invention to a fellow chemist and industrial pioneer, the director of GE's Research Lab, Willis R. Whitney. Whitney had come from MIT in 1900 to GE's Schenectady plant to create the first true research lab in U.S. industry. At the GE Research Lab, which Whitney ran for 32 years, scientists worked alongside engineers and inventors for the first time, and did fundamental research as well as company problem solving. Whitney was a competent colloid chemist, but he soon learned that his real talent was inspiring others. He also understood the purpose of industrial research. "I did not come to Schenectady to create a philanthropic asylum for indigent chemists,"[1] he told the scientists he hired. He inspired them with the value of research, and they left his presence with a strange desire to improve the light bulb—or to find something better to stick on wires to keep the electricity in.

The story of the GE Research Laboratory has been told many times (see Wise, Columbia University Press, 1985[1]). Its results have ranged from very practical improvement of the light bulb to Nobel Prize-winning discoveries in surface chemistry (Irving Langmuir) and superconductive tunneling (Ivar Giaever). Whitney realized, however, that no laboratory could invent more than a small fraction of the improvements needed to keep its company a leader. At his urging GE took a license from Baekeland and began using phenolic resins and compounds for many insulation uses.

Another key GE technologist, Charles Steinmetz, also put insulating materials high on his research wish list. Steinmetz, GE's chief consulting engineer and a visionary, cigar-smoking, German-born Socialist, had proposed the research lab that brought Whitney to Schenectady. In 1912 Steinmetz sent a letter to a second GE lab, the Pittsfield Works Laboratory, urging more work on insulation. Specifically, he urged the laboratory's researchers to explore combinations of polybasic alcohols and polycarboxylic acids in a search for flexible insulating resins.[2] Steinmetz didn't realize he was edging GE toward a promising innovation area of the twentieth century, engineering plastics. That 1912 Steinmetz letter is recognized today as the start of GE's polymer effort. It would turn out to be a marathon, not a sprint.

Electrical Insulation

The GE Works were large multiproduct factories, each product needing many forms of electrical insulation; and the Works management would often choose to manufacture, rather than purchase, the insulations needed. Small copper wires were coated with a baked-on wire enamel, larger cable conductors with an extruded rubber or plastic. The large conductor bars inside motors, generators, and transformers were wrapped with tape insulations of many kinds. Varnished cotton cambric was a common early cable insulation tape, while glass and other cloth tapes, variously bonded, came along in later years. Mica was a key ingredient of both solid insulation forms and tapes, and many varnish types were used to glue the mica flakes together. Insulation varnishes were used for impregnating coils by a dipping, then baking process. Solid insulation spacers were much used in large equipment, and a large quantity of laminates was needed for this purpose. These laminates were made from kraft paper and liquid phenolic resins, formed and cured in heated high-pressure presses. The Schenectady, Pittsfield, and Lynn Works each made laminates, and from them Lynn also fabricated quiet gears and automotive timer components. The Schenectady Works also made insulation varnishes, varnished cambric tapes, and bonded-mica products. The ever-present need for improved electrical insulation stimulated chemistry research in both the Works laboratories and the Research Lab.

Alkyd Resins[3]

In the 1920s an insulation section within the Research Lab explored polyester products from the phthalic anhydride–glycerin reaction, and Roy H. Kienle become a nationally recognized authority on such polymer systems, especially those modified with unsaturated fatty acids. He named these polymers "alkyds," a designation still used today, and he was granted several important patents. The GE Research Lab effort in this polyester field peaked in 1929 at about 10 researchers. The usefulness of alkyd resins in electrical insulation proved limited, but they became and remain a very important resin base for many industrial and consumer paints.

General Electric licensed the alkyd patents broadly and also set up to manufacture alkyd resins and industrial paints, the latter trade named Glyptal, in Schenectady around 1933. The Du Pont line of Dulux enamels for automotive and appliance finishes, for example, used alkyd resins that Du Pont manufactured under a GE license. GE expanded its alkyd resin capacity in the late 1930s, sold alkyd resins externally to paint manufacturers, and eventually achieved a 5–10% market share. GE managed this business along with Schenectady-located varnishes, tapes, and mica products in an organization called Resins and Insulating Materials Department. (RIM).[3]

Molded Plastics: Phenolic Resins and Compounds

After the Baekeland discoveries (patented in 1913), and the subsequent availability of phenolic molding compounds from the Bakelite company, GE

11

products needed plastic molded parts, and several GE plants set up molding shops for their supply. Two of the earliest large-volume parts were the handle and thumbrest for electric irons and molded bases for vacuum tubes. A molding plant was established in the Pittsfield Works, while phenolic resin and compound development in that Works lab under A. McKay Gifford led to manufacturing facilities for improved phenolics. Output of this Pittsfield (phenolic) compound, resin, and varnish plant was also used by the molding and laminating operations in Schenectady, Lynn, and Ft. Wayne, and by the wiring devices (switches and plugs) business of affiliated Monowatt Company in Providence, Rhode Island.

GE combined the growing molded plastics, laminates, and phenolic resin activities of the company into a single Plastics Department in 1930, with sales representation for both internal and external business. Under the leadership of G. Harry Shill, laminates manufacture was consolidated in the Lynn Works in 1932, while molded parts were made in Pittsfield, Meriden, Connecticut, and Ft. Wayne. The growth of plastics-parts molding led to a large new plant on the outskirts of Pittsfield.[4] This molding plant, a $1 million investment in 1937,[5] and, eventually, containing more than 1000 compression presses, was said to be the largest in the United States in the late 1930s. By the 1940s about half the Plastics Department's output of molded parts and laminates was for internal use and half was sold to external industrial customers. Phenolic resin and compound manufacture remained in Building 36 of the main Pittsfield Works, which itself was increasingly dedicated to making large transformers. Wartime demands from the GE laminates and molding plants took the full output of this phenolic resin and compound facility, so these phenolic materials were not sold externally until after World War II.

Plastics Department headquarters and part of the development laboratory moved out of the Pittsfield Works in 1937 to a new location, renamed "One Plastics Avenue," which is still the headquarters location of the GE Plastics world business. The laboratory move to One Plastics Avenue was completed in 1941, and the development staff reached a total of about 75 in 1943.

Under the leadership of Frank D'Alelio and later James J. Pyle, this laboratory yielded some important polymer improvements, including vacuum dehydration of resins and roller grinding of compounds. Phenolic products that were tailored for GE applications proved superior to some available from the larger commercial producers. The GE invention of sulfonated polystyrene materials (D'Alelio) led to significant patents and royalties, though not to manufacture.

In 1940 the lab made important improvements in white laminates to be used for spacer strips between the inner and outer shells of GE refrigerators. The resin developed for these laminates was a novel urea–formaldehyde–melamine mixture, the latter ingredient having recently become available from American Cyanamid. Building further on this technology, the Plastics Lab developed decorative laminates for table and countertop applications, working with a nearby paper company and a printing-roll manufacturer for the top sheet. With four custom designs available, the department announced availability of decorative Textolite in the fall of 1941. But GE withdrew the product line when U.S. war production needs took priority over consumer uses.

Because phenol, a key raw material, promised to be in short supply, the Plastics Department also undertook a wartime project to manufacture phenol in a 15 million pound per year Pittsfield plant. Although the benzene chlorination, caustic hydrolysis, and sodium chloride electrolysis process was used successfully by Dow Chemical, the small, poorly located GE facility had endless technical difficulties and never demonstrated a profitable operation.

Abraham L. Marshall and Research Laboratory Chemistry

By the 1940s GE was dabbling in several fields of chemical manufacture. But these scattered efforts lacked focus. In phenolic materials, GE had some products, but not a lead position. The Bakelite Corporation (later acquired by Union Carbide), the Durez Division of Hooker Chemical, and Monsanto were far ahead. In alkyd resins, Kienle had provided GE with an original invention, but GE was slow to put its own products in the marketplace. Rather than make large quantities for outside sale, it used them internally and licensed patents to others. GE finally lost a challenge to the validity of an alkyd resin patent. Most likely the GE polymer efforts would have diffused away to nothing had not the company found a purpose and a leader at a surprising time, in the depths of the Depression.

The GE Research Lab was in tough shape in 1933. Its first director, Whitney, had suffered a nervous breakdown the year before and retired. His successor, William D. Coolidge, had a great track record as an inventor, having developed both the modern process for making light bulb filaments and the modern X-ray tube. Although ordered to cut the laboratory's staff nearly in half, he was determined not to preside over the dissolution of America's longest running experiment in industrial research. Spurred by this resolve, he secured approval from GE's management to launch, at the depths of the Depression, a couple of efforts aimed at new, innovative areas.

One such area was in chemistry. To lead this effort, Coolidge selected in 1933 a physical chemist who had come to GE from Princeton in 1926, Abraham L. Marshall. It was Marshall who first developed a vision of GE building a chemical business with research leadership. He did not start off with that vision clearly in mind, but did bring to the job a strong background in research, receptiveness to new ideas, and a high degree of determination. His brusqueness could be daunting. His subordinates came to realize that when he often told them, "You (or we) failed miserably," he was simply saying that the final result wasn't outstanding. He nearly drove away the inventor of one key new material by taking one look at it and snorting, "cracked all to hell, isn't it."[2] But his idiosyncrasies were accepted as part of the drive for excellence. Marshall brought in Winton I. Patnode from the Pittsfield Works lab to replace Kienle, who had left the company. Patnode was the first of several Cornell Ph.D.'s to join GE.

In the 1930s, two important Research Lab developments in electrical conductor insulation were completed that enhanced the credibility of Marshall and his associates. Polymers from vinyl chloride had become available from chemical industry sources, but these were brittle and difficult to process. J. Gilbert E. Wright and Moyer M. Safford found materials and methods to plasticize and stabilize vinyl chloride resins so they could be extruded over copper wire as a tough insulation coating, superior to competitive rubber alternates. The GE Bridgeport, Connecticut, wire and cable business trade-named such a vinyl-coated wire "Flamenol" and set up to manufacture it in large quantities. Other wire and cable companies followed suit. GE didn't enter the manufacture and sale of the compounded vinyl polymers; so this materials know-how and a modest patent position were shared and licensed with established vinyl plastic producers and cable manufacturers.

Another insulation challenge was for a tough wire enamel. To be useful for coils in small- and medium-size motors and transformers, the thin enamel film needed electrical integrity over a wide temperature range, plus toughness to endure flexing and abrasion from automatic winding machines in the manufacturing process. Patnode (Research Laboratory), Edward J. Flynn (Schenectady Works Lab), and Marshall undertook a program to create a wire enamel based on Formvar, a tough polymer made by the Canadian Shawinigan Resins Company. After five years of research effort on polymers, solvent combinations, and multiple-coat enameling processes, a vastly improved enameled wire was achieved. "Formex" enameled wire proved so superior to its predecessors that complete redesigns of small motors and transformers became possible and very profitable. Formex enamel was expensive compared to previous alternates, but its performance on wire soon made it the market leader.[3]

Formex wire enamel was formulated in the Schenectady RIM operations and sold to the various GE wire-enameling mills associated with the major Works. The enamel was also sold externally and licenses were granted to other formulators, but General Electric did not take the initiative to manufacture Formvar resin.

Both Marshall and Patnode felt that GE researchers' achievements in alkyd resins, plasticized vinyl chloride, and Formex wire enamel could have (and should have) led to greater participation in chemical manufacturing. And they also were confident that expanded research with man-made polymers would continue to improve electrical insulation and create many other applications.

This confidence was boosted by achievements at Du Pont. There, in the 1930s, Wallace Carothers published brilliant research that helped bring order to the confusing science of making polymers. He confirmed earlier work in Europe showing that polymers were made up of unimaginably long chains of molecules. He explained functionality, and the mechanisms of addition and condensation polymerization by which those chains were put together. For Du Pont, this would result later in the decade in two major new products, Neoprene synthetic rubber and Nylon textile fiber. For technology in general, it provided understanding of what earlier pioneers such as Baekeland and Kienle had done, and inspiration to do more. Two persons who understood the opportunity were Corning Glass Works Research Director Eugene C. Sullivan and GE's Marshall.

Silicone Project in the Research Laboratory

Marshall and Patnode were invited to visit the Corning Glass Works research laboratory in 1938 and were shown samples of glass cloth tape impregnated with a phenyl silicone resin. The Corning research, conducted by J. Franklin Hyde, was exploring the concept of silicone-impregnated glass cloth as an electrical insulation tape. Upon returning to Schenectady, Marshall asked Eugene G. Rochow to apply about half his time to research on silicone polymers.[3]

The original discoverer of silicones was Frederic S. Kipping, an English chemist, who first made them in 1904. He observed "sticky masses" when he hydrolyzed organochlorosilanes and knew that the material had a polymer chain of silicon and oxygen atoms. Kipping was interested in silicon monomers, however, not the sticky polymers. He made many organochlorosilanes using a complicated reaction involving magnesium, called the Grignard reaction. Over thirty years later both Corning and GE researchers were still using that Grignard reaction when they began the effort to develop practical silicone polymers.

Rochow, one of the GE hires of the 1930s, would soon achieve world recognition. Born in Newark, New Jersey, in 1909 he early took up chemistry as a hobby, and distinguished himself in the subject at Cornell, where he stayed to earn a doctorate in inorganic chemistry. He moved directly from there to GE in 1935. His professorial scholarly and reflective manner (traits that later, at Harvard, would win him awards for teaching excellence) hid a more determined side: "Rochow has a certain stubbornness of principle," a supervisor wrote of him. "I believe he is motivated by the right as he sees it and not selfish reasons. His position is that of a moralist in a wicked world."[2]

He was enlisted part time onto Marshall's project, and in 1939 made discoveries of methyl and phenyl silicone resin polymers that resulted in five patents. But as long as any production process needed to consume magnesium in the expensive Grignard reaction, Marshall could see no large future for silicone polymers. He may also have suspected that Corning would team up with Dow, who manufactured magnesium from sea water. So he began phasing out the project and told Rochow to go back to his other task at the GE lab, improving magnesium oxide insulation for heating elements in stoves.

However, Rochow, always a stubborn man, persisted in his work in *methyl* silicone polymers for almost a year of part-time effort, seeking a more direct process for making methylchlorosilanes. He recalled his Cornell assignment as lecture assistant to Alfred Stock, a visiting German professor, and later assisting Stock with diagrams for a book. One Stock experiment had reacted HCl gas with a heated silicon-copper alloy, yielding small amounts of trichlorosilane ($HSiCl_3$). In May 1940, Rochow found that methyl chloride would react with a silicon-copper alloy above 300°C to form methylchlorosilanes. This discovery, later patented, was the defining event for expanding the silicone project of the GE Research Laboratory. Not only had a promising silicone polymer field been identified, but now a process for making methylchlorosilane intermediates had been discovered that should prove much lower cost than the Grignard synthesis employed by investigators up to that time.[6] Rochow would joke in later years that some of the impetus for his direct-process discovery was the administrative nuisance of having to walk to a distant building and persuade an unsympathetic purchasing agent to procure magnesium turnings, in time of shortage, which were needed for Grignard reactions.

Marshall reacted to Rochow's discoveries by assigning new personnel to silicone research, first by transfer from other projects and then by new hiring. He clearly believed that major expansion of silicone research would produce important new polymers and that the research position might form the basis for GE's entry into the business commercially. With full support from Coolidge, GE Vice-President and Research Laboratory Director, and from C. Guy Suits, Assistant Laboratory Director, silicone activity grew from this tiny beginning to a peak of about 35 people by the end of 1944, including many newly recruited Ph.D.'s. In addition to chemists, Marshall started a chemical engineering group in 1942 headed by Charles E. Reed, whom he hired from MIT.[3]

Reed was born in 1913, the son of a Findlay, Ohio, banker. He chose chemical engineering for a career, graduated from Case Institute of Technology, and earned an Sc.D. degree from MIT in 1937. He served on the MIT faculty as assistant professor of chemical engineering, coauthoring with T. K. Sherwood, *Applied Mathematics in Chemical Engineering*, which became a textbook classic. By 1942 he accepted the position of Research Associate at the GE Research Laboratory.[7] Why GE instead of the petroleum or chemical giants? He was impressed when Marshall showed him the field where GE was to build a phenol plant, an indicator of commitment to chemistry. More importantly, he had studied the growing plastics industry and he judged that silicones were a "ground floor opportunity." He found that his background from the world's premier chemical engineering department gave him an edge in GE. MIT professors had helped develop the fluidized-bed process for oil industry catalyst crackers (cat. crackers), a method in which catalyst powders move through vessels and pipes like a flowing liquid. Reed conceived of applying fluidized-bed technology to Rochow's direct process, and his 1943 invention of the fluidized-bed processor for making chlorosilanes,[8] plus separation processes for methylchlorosilane monomers,[9] remains the principal method for making silicone precursors throughout the world.

Silicone chemistry quickly proved an order of magnitude more complex than any previous chemical development at GE. The direct process produced a crude liquid mixture of many methylchlorosilanes that had to be separated by tedious fractional distillation. The chlorosilanes were then hydrolyzed, hydrochloric acid separated, and the low-molecular-weight siloxanes then processed by very different routes through to the higher polymer forms of resins, fluids, and rubber gums. Simplified silicone process equations are shown on p. 38. Rubber gums were then compounded further with fillers and vulcanizing agents added. Each step in the processes toward fluids, resins, and rubber required extensive research and development before any commercial products could be defined. In addition to the Research Lab effort, silicone development aimed at commercialization began at the GE (Schenectady) Resins and Insulating Materials (RIM) organization in 1944. The RIM effort reached about 45 people at the end of that year.

Patnode identified the dimethylsiloxane cyclamers from the trimer through hexamer. He showed that sulfuric acid would cleave siloxane bonds and used the reaction to make linear chain-stopped dimethylsilicone fluids, for which he was granted patents in 1949,[10] and which remain unchanged today in the industry's products.

Cross-linked resin products needed trifunctional as well as difunctional siloxane units of both methyl and phenyl types, so Rochow and William F. Gilliam extended the direct process to yield phenylchlorosilanes from silicon and chlorobenzene with a silver catalyst; or with chlorobenzene, hydrogen chloride, and a copper catalyst. Wright, James Marsden, and Charles Welsh improved the resin-making process by solvent selection and careful posthydrolysis condensation.

Maynard Agens showed that dimethyl silicone hydrolyzates could be further polymerized to a rubbery gum. Wright made higher-molecular-weight gums using ferric chloride catalyst and showed that benzoyl peroxide would cross-link and strengthen a heated methyl silicone gum. Marsden demonstrated that small amounts of methylvinylsiloxane would produce firmer gum cross-linking. With Flynn, he developed useful, heat-vulcanizable rubber by compounding gum with fillers and benzoyl peroxide on rubber milling rolls. Wright compounded silicone gum to an unusual "Bouncing Putty," which later became a universal toy, Silly Putty.

Many GE researchers studied and struggled with the extraordinary complexity and variability of the direct process for making dimethyldichlorosilane in high yield. At one point Rochow absolutely could not reproduce his reaction. The problem was found to have been a shift to electrolytically deposited copper, which, perhaps because cuprous oxide was absent, would not catalyze the process.

The methyl chloride–silicon reaction is exothermic, and higher reaction temperatures favor formation of methyltrichlorosilane and silicon tetrachloride, both undesirable. Thus as reactors became larger, heat transfer for temperature control became an important design consideration and several reactor types were tried: static beds, rotating horizontal drum, fluidized bed, vertical stirred beds, and a stirred, pressurized horizontal reactor. Vertical stirred beds (Jesse E. Sellers and J. L. Davis) and fluidized beds (Reed and J. T. Coe) were chosen for further development. Gilliam and R. N. Meals studied

18

silicon particle-size effects in the direct process, and showed that silicon-copper powder mixtures could replace alloys. R. O. Sauer discovered that catalytic equilibration of chlorosilane mixtures could redistribute methyl groups between the molecules, an important yield-improving process.

Fig. 2-1. GE Research Laboratory early silicone scientist team.
From left, W.J. Scheiber, E.G. Rochow, R.O. Sauer, A.L. Marshall,
W.F. Gilliam, W.I. Patnode, M.M. Sprung, ca. 1941.

Crude methylchlorosilanes separation by fractional distillation was found to require as many as 80–100 theoretical plates and high reflux ratios because of some close boiling points.

(For a complete account of GE silicone research and development, the reader is referred to reference 3.)

Research and product development needs for the chlorosilanes and siloxanes led to small pilot-plant facilities in the Schenectady Research Laboratory and RIM organizations. But serious wartime uses for developmental GE silicones were remarkably few, partly because the company had few sales contacts to focus product development for such a new material. Two significant silicone rubber applications were developed with the help of the Schenectady Works Lab. Gaskets for Navy searchlights in 1943 and for B29 turbosuperchargers in 1944 were the first GE commercial applications of silicone rubber. And a successful chlorosilane vapor treatment of ceramic insulators in high-altitude radios prevented electrical arc-over in humid conditions.

Some application setbacks were encountered: dimethyl silicone fluids failed to lubricate many mechanical systems, so their advantages of flat viscosity-temperature index, high temperature stability, low volatility, and low pour point, went for naught in many potential applications.

At about the same time that Corning's Hyde showed samples to GE's Marshall (1938), he also showed them to Captain Hyman Rickover. Later to become father of the nuclear navy, Rickover was at this time looking for better insulation for submarine motors. He was subsequently able to secure wartime priority for silicones research. This assured Corning and GE of the people and priorities to carry on silicone development during the war. Despite Rickover's drive and the companies' technical successes, no silicone motor insulation ever went to sea in that war.

Silicones Decision Time for GE

The silicone project had reached a size in late 1943 where GE management understanding and decisions were needed for future planning. Competitive silicone activity was visibly accelerating. Corning Glass Works and Dow Chemical had announced a silicones joint venture in late 1942 that would operate initially as the Dow Corning Division of Dow Chemical.[11-13]

Meanwhile, GE Research Lab and RIM personnel made a silicones presentation in December 1943, to the GE Engineering Council, which advised top management on new GE product proposals. RIM received approval in 1944 for a major pilot-plant addition in Schenectady, but did not implement the project, pending consideration of a larger plant commitment.

The RIM proposal to manufacture silicones commercially prompted GE president Gerard Swope (who was recalled to head the company during Charles E. Wilson's leadership of the War Production Board) to ask Zay Jeffries to study the company's chemical-type manufacturing businesses and the silicone project. Jeffries, a metallurgist, had been a consultant on tungsten lamp filament processing to the GE lamp business and had starred as an expert witness during a court challenge to the Coolidge ductile tungsten patent. His advice had also stimulated commercialization of the Research Lab's sintered tungsten carbide technology by forming the Carboloy Corporation, a GE affiliate. He served as its president during difficult depression years and brought it to later profitability. Jeffries' technical and management contributions gave him credibility with Swope and also with Wilson, who later returned from Washington to resume leadership of General Electric.

Jeffries recommended to Swope that GE "plan to expand its chemical activities primarily to provide an outlet for General Electric research in the chemical field..." and, secondly, that a separate Chemical Division be formed.[14]

When wartime secrecy orders covering silicones were lifted, GE held a press conference in November 1944, to announce its research achievements and their potential significance. The *New York Herald Tribune* headline, "Rubber to Last Lifetime Made of Sand and Gas: General Electric Unveils New Product; Foresees Tires to Outlast Trucks" captured the unrealistic expectations engendered by such research announcements. Bouncing Putty, which subsequently never became more than the Silly Putty toy, also got a big play in the press. But while GE in 1944 was talking about future silicone products, Dow Corning had a plant operating to make them in quantity for the war effort. And the leadership in place for this Dow Corning effort was a seasoned management team, fresh from successful experience commercializing Dow Chemical's methyl and ethylcellulose products. Rochow presented a more serious announcement of GE silicone research at a June 1945, Gibson Island High Polymer conference, where he actually demonstrated his direct-process reaction to make methylchlorosilanes.[6] Three Dow Corning researchers also presented papers.

GE Forms a Chemical Division: 1945–1948

Jeffries' recommendation that a Chemical Division be organized was accepted and he became its first Vice-President and General Manager, effective Jan. 1, 1945. The new division, whose headquarters were in Pittsfield, included these departments, products, and plant locations:

Plastics Department. Custom molded plastic parts, with industrial design and mold-making services; molded and extruded silicone rubber parts. Plants at Pittsfield, Meriden (later closed), Taunton, Massachusetts, and later Decatur, Illinois.

Alnico permanent magnets. Plant in Schenectady (later Edmore, Michigan).

Laminates Department. Phenolic–paper sheets and tubes for electrical insulation; white separator strips and inner door panels for refrigerators; Textolite decorative laminate for countertops and furniture surfacing. Plant at Lynn, Massachusetts; moved in 1948 to Coshocton, Ohio.

Chemical Materials Department. Phenolic resins and molding compounds; magnesium oxide powder. (MgO powder, refined from the output of mines in India, was developed in the 1920s by the Research Lab and the Pittsfield Works as insulation for red-hot heating wire inside metal-sheathed heating elements of electric irons, ranges, and other heating appliances.)

Plants within the Pittsfield Works. Phenolic resin manufacture later added at Coshocton, Ohio.

Phenol (internal use). Plant at Pittsfield, Massachusetts.

Resins and Insulating Materials Department. Alkyd resins, industrial paints, insulating varnishes, wire enamels, bonded mica products, insulating tapes. Plants in various locations within Schenectady, New York, Works. RIM also built a small alkyd resin plant at Anaheim, California, in 1948.

Silicone resins, fluids and fluid emulsions, rubber gum and compounds. Greenfield plant under construction in 1945 at Waterford, New York. Production began in 1947.

(Note: In 1949 the RIM organization was disbanded. Silicones, alkyd resins, wire enamels, and varnishes were assigned to the Chemical Materials Department, tapes and mica products went to the Laminates Department.)

Of the original division product lines, molded plastic parts had the largest sales, with laminates next. These products were not strictly "chemical," but were fabricated and semifinished materials. The early division products, except for silicones, had a management structure that had been in place for several years. Silicones was the major new plant investment, and here GE was starting with few well-defined products, few established customers, and no experience in a marketing and technical challenge of this complexity. The RIM organization, with responsibility for silicones, had effective selling contacts within GE, and externally with the paint industry, but it totally lacked representation with other industrial customers.

The 1945 Chemical Division set up a pooled selling organization handling all products, patterned after GE's successful Apparatus Sales Division. Also reporting at Division level were a new product development laboratory (an offshoot of the former Plastics Laboratory) and staff process engineering components in Pittsfield and Waterford reporting to Reed.

Difficult Early Years

General Electric's newly formed Chemical Division experienced very difficult early years. Its large position in thermosetting molded parts was continually eroded by customer shifts to newer thermoplastics and by the trend for large customers, including GE operations, to install captive "in-line molding" capacity. New plastics and molding technologies became broadly available from materials producers and equipment manufacturers, so technical advantage by a custom parts molder was difficult to achieve.

Laminates management had fast-growing markets and good technology, but soon had to cope with the confusion of a rapid move to a new plant, in effect being pushed out of the Lynn Works by the increasing space needs of the new GE aircraft jet engine business. Decorative laminates, which the Pittsfield development lab had pioneered, simply had too little manufacturing capacity in the early years, and GE forever forfeited market leadership to Formica Corporation, later a part of American Cyanamid.

Managers of GE phenolic resins and molding compounds were seeking external business by product innovation in competition with industry leaders such as Union Carbide (Bakelite), Durez, and Monsanto, all of whom were basic in making phenol. GE shut down its miniature, unsuccessful phenol plant in 1949.

RIM alkyd resin products and insulation materials were in good demand, though the paint resin market had many competitors and large paint companies were beginning to make alkyd resins for their own use.

In silicones, major patent issues between GE and Dow Corning had surfaced in late 1944 as wartime secrecy was lifted and the Patent Office declared interferences between inventors' claims from each organization. Although the Rochow direct-process invention was not in interference and was issued in August 1945, other GE filings, notably in siloxane product areas, were in dispute with scientists of Dow Corning or its parent organizations. GE researchers believed that their patent position would emerge stronger after the interference resolutions, but this would take years to adjudicate. Jeffries believed that both parties and certainly the silicone markets would benefit through cross-licensing rather than protracted patent-by-patent litigation. Dow Corning management agreed, and each company gained royalty-free license rights under such patents as might issue from each other's early filings. Rochow had protested this decision to Marshall, to Suits, and even to Jeffries, and he was very disappointed that competition had gained a free license to the process he had invented.[6]

GE silicone sales in 1948 were less than $1 million, a small fraction of the $5 million initial plant cost. Tardy training and fielding of a silicone sales force was a key problem, which in turn made the GE silicone product offerings slow to reach the competitive standards established by Dow Corning.

GE Chemical Division sales in 1948 were around $50 million, about half of these for internal GE use. U.S. chemical industry leader sales that year were: DuPont, $969 million, Union Carbide $632 million, and Dow Chemical, $171 million. Earnings of the GE division were negligible, drawn down by losses in silicones and marginal results elsewhere. Ralph J. Cordiner, then GE Vice-President—Assistant to the President, counseled the division management to "Consolidate your position," meaning solve the profit problems before proposing any more new ventures. Corporate GE had a low tolerance for unprofitable or marginal products.

But also in 1948 the company was building a splendid new home for the Research Laboratory, overlooking the Mohawk River, five miles northeast of its old location within the Schenectady Works. Here Marshall's chemical researchers would have fine new labs and support services, as well as a separate chemical engineering pilot-plant building.

Early Years' Summary: 1900–1948

Chemical competence in General Electric grew naturally from the company's earliest beginnings because electrical products required insulations in many forms; and as moldable plastics became available GE products needed parts made from them. Managers in the large GE Works preferred to make, rather than buy, electrical insulation and molded parts, and these manufactures drew chemical competence to the large plants; especially in Pittsfield (phenolic resins, molding compounds, and molded parts), Lynn (laminates), and Schenectady (alkyd resins, insulating varnishes, wire enamel).

Thus, formation of a GE Chemical Division in 1945 was a natural evolution from several bottom-up management initiatives; it was not a top-down management decision to enter the chemical business. The silicone opportunity was a forcing factor leading to the decision. The company approved the silicones plant venture, despite an obvious lack of proven products, markets, and customers, in response to vigorous urging by the Research Lab leaders, Coolidge, Suits, Marshall, and Patnode, and by the Resins and Insulating Materials management. These men lobbied for GE to manufacture silicones rather than ceding the field to others by broad licensing.

GE managers of electrical equipment businesses were opposed to the Chemical Division formation and its later expansions, because their large chemical industry equipment customers, such as Du Pont and Union Carbide, didn't care for GE competition in their industry.

If GE had been doing formal strategic planning in 1948, the Chemical Division would most likely not have been chosen for aggressive investment in the future. Custom plastics molding was being upset by technology change and customer in-line molding; decorative laminates had fallen far behind Formica; alkyd resins market share was low and margins were declining; GE was not basic in two key raw materials, phenol and phthalic anhydride; and silicones product and market leadership had been captured by the fast-moving management team at Dow Corning. However, GE did have the technical position in silicones provided by Rochow's invention and the subsequent broad research program. Without this there would have been little support for expanded chemical industry participation. With it, GE felt justified in the large plant investment, and would now find out if, after 20 years of modest initiatives, it could succeed in the chemical industry. But absent a silicone business success there would be little company support for continuing.

Fig. 2-2. In the early 1940s GE's Research Laboratory occupied parts of Buildings 5 (left) and 37 (right) at the Schenectady Works. Marshall's Chemistry Section was then on floor 5 of Bldg. 37, with the methylchlorosilane pilot plant on the 6th floor, left corner.

Fig. 2-3. GE's Corporate Research and Development Center, 1980s.

CHAPTER 3

GE Silicones:
1940–1964

From Shaky Start to
Successful Business

In the years 1945–1949 GE nearly fell flat on its face in the silicones business. GE had some important technology and experience in simpler chemical businesses, but it lacked other prerequisites for silicone success: sufficiently experienced general management, marketing skills, customer connections, applications expertise, manufacturing experience, and chemical engineering know-how. It faced a rival that had all of those things: Dow Corning combined the silicone research position of Corning Glass Works with the chemical business experience, personnel, and facilities of Dow Chemical.

GE did not seriously challenge Dow Corning, which half a century later remains number one in silicone sales in the world by a wide margin. GE did, however, succeed in establishing a profitable silicones business. It succeeded for several reasons: the discoveries by Eugene G. Rochow, Winton I. Patnode, and many other GE scientists; the ability of some GE chemical engineers and managers to learn on the job; the silicone growth market for both military and civilian uses; the technical sophistication and capital intensity of this chemical manufacture, which created a high competitive entry hurdle; and the small size of the business compared to others in GE, which encouraged corporate headquarters to either be patient or overlook it.

Dow Corning

The research beginnings of the Corning and Dow Chemical partnership expanded concurrently with those at GE. GE's research program became evident to Corning in the fall of 1940, when Rochow and Gilliam and other GE chemists presented papers at an ACS meeting. J. Franklin Hyde, at Corning, also published his work shortly after this. Corning was also sponsoring siloxane research at the Mellon Institute under Rob Roy MacGregor and Earl L. Warrick. Formation of Dow Corning Corporation combined the two company efforts in 1943. The Corning/Dow research programs and General Electric's each made independent discoveries of the early routes to silicone resins, fluids, and rubber, but the Dow Corning management team achieved significant sales and production much faster than GE. Silicone market development in the early years is essentially the Dow Corning story, well told in Warrick's *Forty Years of Firsts,*[1] MacGregor's *Silicones and their Uses,*[2] and in oral histories by Hyde and Warrick.[3,4]

Corning's Research Director Eugene C. Sullivan had sought help from Dow to scale up Grignard reactions to make an intermediate needed for Hyde's resin. Technical liaison between the two companies grew, and early in 1942 William R. Collings, manager of Dow's Cellulose Products Division, was asked to take overall charge of the Dow silicone effort, using the personnel and facilities of his division. Collings and his experienced staff swung into silicone research, product development, pilot production, and sales. Shailer L. Bass (Research), Howard N. Fenn (Production), Melvin J. Hunter and Arthur J. Barry (Research), Toivo A. Kauppi (Technical Service), Olin D. Blessing (Sales), and others in Cellulose Products joined the silicone program within Dow and later with the new company. The Dow Corning venture was announced in September 1942, and the new company, incorporated early in 1943, soon outgrew its pilot plant and broke ground for a new plant, which began production in May 1944. By this strategy Corning gained an experienced management team and substantial Dow Chemical facilities in support of its silicone effort. Dow was obviously impressed with the potential for silicone polymers, as well as attracted by use of its magnesium for Grignard processes, then needed to make the organochlorosilane intermediates.

Corning's Mellon research had prepared dimethylsilicone fluids and some early silicone rubber, Hyde's resin for electrical insulation was urgently wanted by the Navy, but the product in greatest demand for the war effort turned out to be a grease sealant for aircraft ignition harnesses. This grease was critically needed to prevent engine failure at high altitudes from corona formation and arc-over at the spark plugs. Bass's contact with a Wright Field technical committee identified the need in early 1942, and he developed a successful formulation from a Mellon dimethylsilicone fluid and a silica filler. This silicone grease, named DC 4 Dielectric Compound, solved the serious failure problem and was superior to organic formulations in resistivity, dielectric strength, arc resistance, flash point, water absorption, and useful temperature range. Military demand for this grease and for 990A resin required maximum pilot-plant production effort, and supported wartime priority allocations for the materials needed in new plant construction and the use of scarce magnesium. Sales for these and other military applications allowed Dow Corning to report a profit in its first year of operation. When wartime silicone demand halted abruptly in late 1945, the Dow Corning team quickly reoriented to peacetime market development, searching for applications in which unusual silicone properties would justify initial prices in the $5–6 per pound range.

Fortuitously, a Dow Corning sales engineer's 1945 presentation to U.S. Rubber Company's Tire Division led to the successful trial of a silicone fluid-water emulsion-mold spray to effect tire release from molds without sticking. Reduction in tire rejects justified the silicone cost many times over, and this one customer's requirements equaled Dow Corning's capacity at the time to make dimethylsilicone fluid. Dow Corning also soon demonstrated a baked resin coating for bread pans, called Pan-Glaze, that was a cost improvement for the baking industry, replacing lard lubricant then used on every bake cycle. The rapid success of these two large civilian applications gave the Dow Corning organization great confidence in the growth ahead.

Dow Corning's "Silastic" brand silicone rubber compounds dramatically improved in strength and in compression-set, and several rubber fabricator customers became specialists in molded and extruded silicone parts. Textile water-repellents were successfully formulated, as were nonstick coatings for release paper. Electrical resin applications grew slowly, first in heavy-duty traction motors, and later throughout a new high-temperature insulation system. As customer inquiries poured in from all these markets, Dow Corning rapidly fielded a trained sales force across the United States. The field sales force, including local support, reached 68 by 1953 (GE had less than 20 at that time), and were backed up in Midland by technical experts in product and application.

During this period, specialized medical applications, such as fabricated rubber parts for body implants, were developed by a separate headquarters group, which was staffed and equipped to serve medical research and development.

All Dow Corning pilot production and its first plant used Grignard processes to make the chlorosilane intermediates; and, according to Warrick, the company's Grignard facility became the largest in the world. The subsequent expansions for methylchlorosilanes used the Rochow direct process, following the cross-licensing of early patents negotiated between Dow Corning and GE in 1945. Dow Corning used rotating kiln type reactors early, later followed by fluidized-bed systems. Phenylchlorosilanes were first made by Grignard and later by a Barry-invented process in which trichlorosilane reacted with benzene.

Dow Corning moved early to become a factor in world markets. Bass negotiated agreements for equity participation and patent licensing with U.K. and French companies, distributors were set up to serve other major nations, and international sales and technical service was given separate management.

Collings' leadership of Dow Corning spanned 1943 to 1962, first as vice-president and general manager (Sullivan was Dow Corning's first president), and then as president. Sales grew to $54 million while prices fell to the $3–4 range, and the company was highly profitable. Its nearest competitor, GE, had less than 40% of this volume. Dow Corning clearly deserved its slogan, *First in Silicones.*

General Electric Silicones[5]

GE's silicone plant at Waterford, New York, on the Hudson River twenty miles northeast of Schenectady, started up methylchlorosilane production in mid-1947 and siloxane products later that same year. Multiple, 6-inch diameter, vertical stirred-bed reactors, which had demonstrated the best yields of dimethyldichlorosilane, were used for the direct process. Phenylchlorosilanes were purchased from Dow Corning.

Initial finished products included resins for mica bonding and for high-temperature paints, dimethylsiloxane fluids in a range of viscosities, and a rubber gum sold to the GE Plastics Department, which compounded it and then formed gaskets to seal small fluid-filled GE capacitors. A rubber paste developed to seal the chamber in GE steam irons proved a substantial cost improvement over brazing. But overall sales growth was painfully slow. As Waterford plant inventories built up in 1949, methylchlorosilane production was halted for several months, though siloxane production and sales continued. Major management changes were needed, and these were initiated across the board.

Harold F. Smiddy, who would later assist Ralph J. Cordiner in the latter's reorganization of General Electric, the building of GE's Management Institute at Crotonville, New York, and the "four blue-books" manager course (Plan, Organize, Integrate, and Measure) became vice-president and general manager of the Chemical Division and implemented a small task force review of the silicone situation. The knowledgeable group noted that the low sales level was the key profitability problem, and that this stemmed from the lack of a specialized field sales force and a lagging product development effort. Task force member and former silicone researcher Patnode, by then heading an atomic laboratory for GE in Richland, Washington, summed up his view to me, "Not enough dirty glassware" (in the product development labs). By this time the silicones effort at the GE Research Lab was phasing down and only a few of the RIM silicone development personnel had moved from Schenectady to Waterford.

Despite the obvious problems, Smiddy, as had Zay Jeffries before him, strongly supported the silicone effort. He brought Cordiner to Waterford in 1949, where he and Charles E. Reed acknowledged the poor business situation, outlined plans to solve problems, and stated their confidence that silicones still represented a good opportunity for GE, probably the best opportunity within the Chemical Division. Waterford employees were fearful that the outcome of this meeting might be a business shutdown; but Cordiner became a supporter of the silicone effort, though he was impatient for the program to reach profitability.

The division pooled-sales structure was dissolved and split into product specialty groups. Silicone sales specialists were trained and put in the field, reporting to an industry-experienced sales manager at Waterford. A combined salary-plus-incentive compensation for field sales representatives was put in place across the division, a rarity at that time in the chemical industry, and all sales reps were required to write call reports. Several experienced chemical executives were recruited for GE's Chemical Division, both at Pittsfield headquarters and at the silicone plant.

James W. Raynolds, an executive with a broad chemical industry background, was brought in from the Sun Chemical Company as Assistant General Manager for Silicones and located at Waterford. Raynolds's superior in Pittsfield also presided over phenolic products, alkyd resins, and magnesium oxide. While considering the GE position, Raynolds contacted his acquaintance, Willard H. Dow, President of Dow Chemical, for his view of the GE silicone situation. Dow suggested that Raynolds could certainly help, as the GE effort was scarcely noticeable to Dow Corning at that time. Raynolds quoted Dow as saying, "We don't know they're alive." Raynolds joined GE in 1948, six years after Dow's comparable general management assignment of Collings to lead the Dow Corning silicone effort.

Laboratories Before Offices

Faced with limited resources and many different product-development opportunities, Raynolds made a clear choice to expand silicone gum and rubber-compound development, an arena in which he had confidence from his rubber industry experience. New-product-development laboratory space was created out of former offices, while sales and administrative personnel, including the general manager, moved into the farmhouse, construction sheds, and basement facilities on the property. It wasn't good form in General Electric for a loss business to spend on office improvements. Raynolds also resisted suggestions from other industry-experienced hires on the Pittsfield division staff that GE's silicone business concentrate on selling methylchlorosilane intermediates rather than developing the extraordinary diversity of siloxane polymers needed to compete in the finished-product market.

Fortunately for GE's growing, though unprofitable silicone business, RIM management had added a substantial warehouse on the Waterford property in 1949, first to provide storage for imported mica and later to house a planned move of RIM Schenectady manufacturing. As time passed, RIM could never justify the 20-mile move to Waterford, so this warehouse gradually filled with silicone-gum kettles and silicone-rubber compounding equipment, plus a silicones shipping dock and manufacturing office space.

Beginning with a mold release emulsion for tires the GE silicone line slowly reached competitive levels, product by product. Industrial customers prefer more than one source for purchased materials, so the GE field sales reps used the second-source argument effectively with established users. GE fluids qualified with the major manufacturers of car and furniture polish in time to catch the growth wave of these new, easy-to-use polish formulations in the 1950s. These polish applications depend on the different surface properties of silicone fluids and the traditional waxes, the combination yielding a high-gloss, long-lasting coating with minimum polishing effort.

Silicone Rubber

Rubber markets were a major challenge because competition had moved ahead in compound quality beyond anything that could be achieved with GE's early gum. Reed mobilized extra effort at the New Product Lab in Pittsfield, where Glennard Lucas demonstrated a breakthrough high-molecular-weight gum starting from high-purity dimethylsiloxane tetramer and following the Hyde KOH catalyst polymerization practice. This polymer (SE 76) quickly became the mainstream gum for GE compound development. GE continued a "gum approach" marketing strategy, hoping to interest rubber fabricators in performing the compounding step with fillers and catalysts, as was industry practice with natural and synthetic rubber polymers. This didn't catch on in silicones for several reasons. Silicone-rubber compounding is very sensitive to impurities, including carbon black, and so requires separate milling equipment. Also, fabricators in this specialty field couldn't match compound improvements being made by the primary silicone producers. Years later the silicone industry would successfully offer "reinforced gum bases," in which some reinforcing filler plus many critical and proprietary additives were mixed with the gum polymer. Additional fillers, colors, and the peroxide curing agents could then be added by the fabricator or by custom compounders.

Cordiner Reorganizes General Electric

Ralph J. Cordiner, who succeeded Charles E. Wilson as chief executive of General Electric, reorganized the company beginning in 1951.[6] His principles included clearly defined business product scopes and profit responsibility assigned to a general manager, who had control over the resources needed to conduct the business, and who would manage for acceptable results without interference from above or alongside. Internal product sales were to be priced at external market levels, and departments selling internally were expected to compete successfully outside the company as well. Geographical as well as managerial decentralization was encouraged, so that in the 1950s many product lines from GE's multiproduct Works moved to new plant locations.

Top to bottom, layers of management were reduced by decree to seven. Strict organization nomenclature was established: *Department* was the smallest full-function profit-and-loss (P&L) unit, and functions within a department (e.g., Manufacturing or Marketing) were called *Sections*; *Divisions* consisted of several Departments, and *Groups* were then the largest operating components. The Cordiner reorganization also abolished all formal *Committees* and all job titles containing the words *Assistant* or *Assistant to.*

This program caused major restructuring of GE's electrical equipment businesses and, in the Chemical Division, prompted designation of silicones as a separate department. The one-profit, one-manager philosophy of the Cordiner reorganization affected organization evolution of the GE Chemical Division, and the company, for years to come.

Reed, who had been in charge of Division technical staff components, moved from Pittsfield to Waterford in 1952, becoming the first Silicone Products Department general manager, while Raynolds became head of division marketing staff in Pittsfield.

Sales Growth by New Products

Failure of GE's silicone-rubber "gum approach" in the marketplace convinced management to follow Dow Corning's specialized rubber compounds example, and the GE program became successful in a few years' time. Silicone-rubber parts used on military aircraft expanded rapidly with the Korean and then the Cold War. GE added a diphenylsiloxane modified gum and compounds to improve low-temperature flexibility from - 65° to -120° F. The Navy rewrote power and communications cable specifications to take advantage of silicone-rubber heat stability and nonconducting ash under catastrophic fire conditions. And nonmilitary applications from oven gaskets to baby-bottle nipples were growing. GE gained customer preference in several market niches and earned a reputation as a supplier of top-quality silicone-rubber. The improvements in silicone rubber resulting from research by all producers and from the availability of fine-particle fumed silica fillers from Degussa Company was dramatic, as noted below in Figure 3-1.[7] Dow Corning discovered a technology to solve a serious "crepe hardening," or "structure" problem associated with compounds containing these very fine fumed silica fillers. All producers later added small amounts (ca. 0.1%) of methylvinylsiloxane to the rubber polymers, which reduced the required amount of peroxide cure agent substantially and improved rubber parts resistance to compression-set.[8]

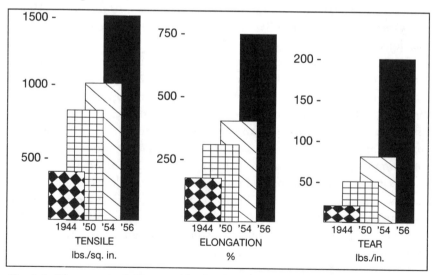

Fig. 3-1. The strength of silicone rubber has been greatly increased.
(From R.N. Meals and F.M. Lewis, Silicones, Reinhold, New York, p. 240 (1959). Reprinted with permission.

Combined with silicone rubber's extraordinary resistance to sunlight and oxidation damage, plus high-temperature stability and good compression-set, these strength gains and some price declines stimulated the wide growth in nonmilitary applications that has continued to the present day. Dow Corning introduced a trifluoropropyl siloxane gum modification to reduce swelling of cured parts in solvents. GE tried, but failed, to introduce a satisfactory nitrile-modified silicone rubber (NSR) to compete with fluorinated product; and later took a license under the Dow Corning patents.[8]

GE continued to improve and expand its resin product line, with new products for paint makers, for masonry water-repellents, and for expanded use as electrical varnishes. Silicone transformer varnish applications at GE plants were successfully wrested from the competition. Fluids development increased the viscosities available to 60,000 cps and a GE fluid called Versilube F50 dramatically improved the lubricity of a dimethylsilicone by incorporating a small percentage of chlorinated phenylsiloxane. Specialties such as antifoams and greases were added to the product scope.

In 1953 I began the analysis shown in Figure 3-2, which quantified the contribution of new products to the department's growth. The business was still in the red, and we frequently had to justify the large R&D programs. The contribution of new product development to GE silicone sales growth became immediately apparent by graphing the sales segments by year of product introduction.[9] I also found that the total expense of new product R&D was returned as a rapid payback from the margin on new product sales. As the years went on, the top five sales elements, comprising the most recent products, contributed 42% to 54% of annual sales over a ten-year period. These analyses confirmed the essential strategy of investing heavily in product research and development.

Our sales growth required a major plant expansion in 1952, including larger methylchlorosilane facilities, this time with 24-inch-diameter, vertical stirred-bed reactors designed by Austen W. Boyd, more rubber gum and compounding capacity, and backward integration to make methyl chloride from methanol and the hydrogen chloride hydrolysis byproduct. Technology for this latter process was purchased from Du Pont, our methyl chloride supplier. Simplified process equations for making silicone fluids and rubber gum are shown on page 38.

In spite of growing sales and diminishing annual losses, the company's top management was occasionally impatient with the progress. Reed tells of receiving a verbal directive in 1954 to "Break even this year, or else!"[10] He and the department were able to comply, without cutting R&D, thanks to a cost reduction from starting up a phenylchlorosilane direct process unit and

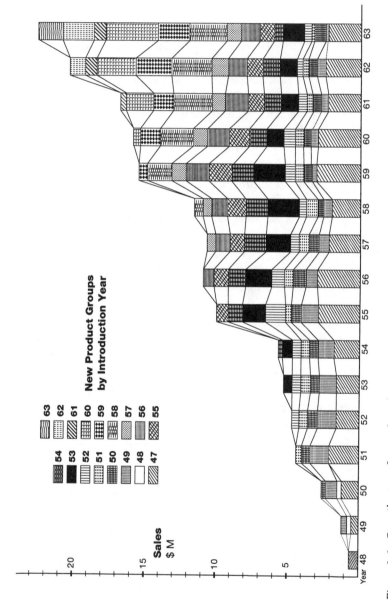

Figure 3-2. Contribution of annual product groups to GE silicone sales. (Adapted from J. T. Coe, "Measuring the Success of R&D Innovations," National Productivity Review, Summer 1982, pps. 331, 333)

ending purchases from competition. Break-even came seven years after the Waterford plant opened.

Sales surged in 1955 and department earnings showed the promise that we had been forecasting for so long. Reed then gained approval for a fine new silicones research and development laboratory, plus conversion of older lab space to serve the growing technical service staff. Laboratories still had priority over offices, though it was planned to move the general manager's office from a farmhouse to the new R&D building. But in early 1959, before having a chance to savor his new office, Reed was promoted to manage the larger Metallurgical Products Department in Detroit; and I succeeded him as silicones general manager.

International Sales and Licensing

The GE Company's international sales, distribution decisions, and patent licensing had long been the responsibility of the International General Electric (IGE) Company, a company component headquartered in New York. GE Chemical Division managers were surprised to learn that IGE had made early licensing commitments for silicone patents abroad to Imperial Chemical Industries (U.K.), Rhone-Poulenc (France), and Farbenfabriken Bayer (Germany). No rights had been negotiated with these licensees for GE joint ownership of the silicone projects. Later, Japan patent licenses were negotiated with Shin Etsu and Toshiba. So while the GE silicone business received significant licensing and know-how exchange royalties for many years, sales participation in Europe and Japan was limited to distributors set up by IGE, which faced strong competition from the GE licensees. GE's silicone sales outside the United States remained in single-digit percentages of the total for many years, far less than that achieved by competitor Dow Corning.

Union Carbide Enters the Silicone Field

While Dow Corning and GE silicone sales were growing rapidly, Union Carbide (UC) was preparing to become the third U.S. producer. UC had an insider's look at silicone industry growth because their Electrometallurgical Division was the major supplier of elemental silicon to both Dow Corning and GE. UC's Linde Division obtained patent licenses from GE and from Dow Corning, conducted research at Tonawanda, New York, and put an integrated plant on-stream at Sistersville, West Virginia, in 1956–1957. The UC Silicones Division later separated from Linde and grew successfully, but around 1965 was joined, for general management and marketing, with other Carbide chemical products.

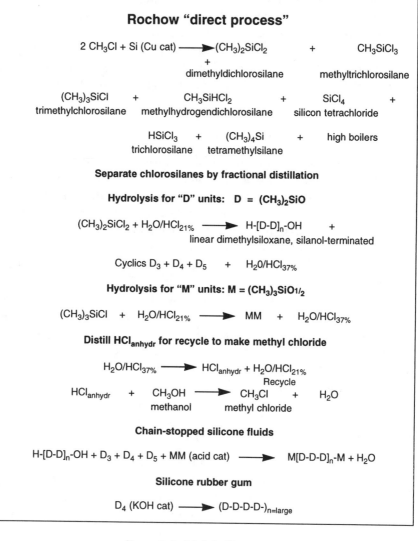

Figure 3-3. Methyl silicone processes.

UC not only competed in established silicone markets, but pioneered new silicone applications such as surface treatment of glass fibers with alkoxyaminopropylsilane, and with a novel dimethylsilicone polyether block-copolymer, marketed a bubble-size-control additive for urethane foam production. This latter application was a major success for UC worldwide. GE took a license from Union Carbide and competed in the polyurethane foam additive (PUFA) market.

Market Growth and Diversity Continues

The silicone market and GE sales continued to grow on all product fronts in the 1960s. No single application, product, or customer dominated the market, and each of several hundred product grades needed tight quality control to satisfy customer requirements and justify silicone's premium prices. While this market complexity made competitive entry more difficult, production scheduling and quality control of the diverse offerings was challenging.

GE had a serious identity problem, because few potential new-application customers were aware it made silicones. GE slowly improved its silicone supplier recognition by direct-mail advertising and publicity, which together stimulated the flow of more than 10,000 samples a year. As product-line

Fig. 3-4. J.T. Coe, E.G. Rochow, C.E. Reed in Waterford lab, ca. 1962

improvements came along, the field sales force proved a critical factor in successful sales and customer development. GE had been slow to increase the sales personnel resource in early years, but expanded the force rapidly in the late 1950s, adding a district sales-manager structure and later silicone-rubber specialists for that growing product line. Also to promote silicone rubber, GE assigned end-user specialists in the field to call on design engineers of aircraft, cars, and many other rubber end users. The company reorganized the technical service engineers and market specialists at Waterford into market development units, which guided product planning and supported the field salesforce in closing new applications. R&D personnel were also mobilized when a product modification was needed for a customer application approval.

The mix of growing GE sales required continuous plant-capacity additions for methylchlorosilanes and for the various siloxanes. GE installed its first commercial fluid-bed reactor for the direct process in 1962, building on development by Boyd and Hart K. Lichtenwalner, incorporating inventions by James M. Dotson that made it a semicontinuous operation. This and other efficiency advances for making dimethyldichlorosilane proved crucial to successful economics of the direct process. While Dow Corning had built outside Midland, Michigan, and at two new locations in Kentucky, GE chose to expand at its Waterford, New York, location.

Room-Temperature Vulcanizing Silicone Rubber

GE optimized the technology for a new kind of silicone rubber in the late 1950s. Up to this time silicone-rubber molded or extruded parts were vulcanized by heat. A few formulations had been developed that cured at room temperature after mixing with a catalyst, but market response had not been overwhelming. GE chemist Charles A. Berridge developed a flowable compound from a silanol end-group gum polymer, which, when mixed with an organotin curing catalyst in the two-package system, cured at room temperature and showed some excellent rubber properties. Enthusiastic acceptance of RTV 60 as an airframe sealant and the promise of other formulations as construction sealants, encouraged GE to back this program with a major development and marketing effort. In a few months it also became possible to offer *one-part* room-temperature vulcanizing (RTV) products, in which the room-temperature cure, invented by Rhone-Poulenc, used atmospheric moisture to trigger a catalyst. The one-part RTVs could be sold in caulking cartridges or in tubes, as for consumer bathtub sealants. RTV technology products were the first silicones with which GE entered the market as soon as Dow Corning. It was a "run to daylight" opportunity, and the company geared up to take full advantage of it.

Competent sealant manufacturers were already established in these markets, for example, compounding polysulfide polymers from Thiokol Corporation, and these companies were pressing GE to sell a gum for them to compound and market. Our management team was divided on this policy issue; but I decided at this time we would perform the compounding, packaging, and marketing ourselves, rather than offer the sale of gum. Extending the GE RTV product scope toward the customer involved major complexities in product and market development, and the organization had to acquire these skills. The company found, for example, that its engineers were unfamiliar with equipment for rapidly filling caulking cartridges and tooth-paste tubes with viscous RTV compounds. That packaging problem was solved by hiring Joseph Abbott from GE Lighting, who understood high-speed machinery from his experience making light bulbs.

GE's RTV product line performed equally well in the market with Dow Corning, and the RTV silicone rubbers grew to become a very important part of the silicone industry.

Fourth U.S. Competitor: Stauffer Chemical

As the early GE and Dow Corning patents began to expire, Stauffer Chemical was urged by leaders of Anderson Laboratories, with whom they had an organometallic joint venture, to enter the silicone business. Anderson had been interested in the methylchlorosilane direct process and silicones for many years and had conducted some pilot-plant work. Observing that Dow Corning, General Electric, and Union Carbide were all apparently successful in the early 1960s, Stauffer built an integrated silicone plant at Adrian, Michigan. This market entry was not successful, probably because Stauffer underestimated the technical complexity of processes and products and the customer support level needed to compete effectively.

Directions Not Chosen

Although it was known that Dow Corning had integrated backward to make the silicon metal from quartz sand and carbon in an electric furnace, GE never added this process to its flow sheet. Also, GE silicone management elected to stay out of silicone medical devices, believing that breast implants, in particular, would be viewed by company top management as completely inappropriate for General Electric. Silicones for application to clothing in dry-cleaning establishments were thoroughly investigated, but did not produce the hoped-for customer values. We marketed liquid RTV silicone rubber seriously in several roofing system applications, starting with the GE building at the 1963 New York World's Fair, but the systems chosen then did not prove satisfactory.

Respectability Within General Electric and Lessons Learned

By the mid-1960s the GE silicone business achieved returns on sales and investment that were respectable within the company, and GE chemical managers had gained the credibility essential for continuing approval of plant expansions and new ventures to come. Integrated chemical business-es like silicones are capital-intensive, requiring much larger plant investment relative to sales than typical electrical manufactures. So the GE chemical managers aimed for higher-than-company-average return on sales in order to achieve satisfactory return on investment. And for a given sales increase the chemical businesses needed larger or more frequent appropriation approvals than their electrical counterparts.

GE thus became a successful competitor in silicones over a long period. Competence grew in all functions within the GE organization, which held its number-two market position while turning back serious challenges from Union Carbide and Stauffer Chemical. By having lost position in the early years to the superior speed and effectiveness of Dow Corning's effort, GE people learned many lessons from the silicone experience. GE would not have entered silicones manufacture without Rochow's direct-process discovery, whereas Dow Corning pushed into plant construction with only the expensive Grignard process to make intermediates. Confidence in future developments plus military demand for some key products spurred Dow Corning to get into production quickly, using a less-than-optimum inter-mediates process. Speed to reach and supply the market was more impor-tant to Dow Corning than short-term materials efficiency.

The direct-process patents were a valuable GE bargaining tool and an important royalty generator in the early years, but did not give the compa-ny exclusive cost advantages. GE was slow to assign focused general man-agement and slow to field a silicone sales force. The company learned that prompt customer feedback is essential for a successful product development effort. Market share in a new technology industry is established early, includ-ing in those markets outside the United States. GE neglected to reach for silicone ownership position overseas when its patent and know-how position was most favorable.

If lessons from the silicone experience were to benefit GE in the future, someone had to carry the management knowledge forward. GE was for-tunate that Reed and several of his associates, plus their successors, applied the experience brilliantly while commercializing future GE chemical discoveries.

* * *

We now interrupt the GE silicone story to summarize below the overall Chemical Division picture in the late 1950s. Then in Chapter 4 we track an innovation that eluded GE. In Chapters 5, 6, and 7 we describe the commercialization of three other Research Laboratory discoveries. My silicone association ended in 1964 when I accepted a promotion to the GE corporate marketing staff.

View from Chemical and Metallurgical Division Level: 1955–1960

By the mid-1950s, Robert L. Gibson, Vice-President and Division General Manager, could point with pride to GE Silicones' profitability achievement. Industrial laminates were doing well and a modest position had been gained in decorative laminates. Alkyd resins were declining in profitability, but phenolic resins and molding compounds had developed a profitable specialty niche in the external market. The plastic molded parts business continued to decline. The profitable Carboloy Corporation (tungsten carbide tools), formerly a GE affiliated company, was made part of corporate GE, re-named Metallurgical Products Department and in 1951 was joined with the chemical businesses of this history to form the Chemical and Metallurgical Division. Division sales now amounted to about 3% of total GE and were more than 70% external to the company.

Gibson was not a technical man, having come up through GE's corporate advertising organization, and while he espoused no particular product strategy for division growth, he had confidence that expanded chemical research at the Corporate Research and Development Center (CR & D) would provide some commercial opportunities. Gibson rearranged staff components in Pittsfield to form a Chemical Development Operation (CDO) which would pursue suitable new ventures with all the people and resources needed for research, development, pilot-plant production, and preliminary marketing. CDO costs were not allocated to the operating departments, but were kept as a line item at division level. Gibson would defend that cost as necessary to incubate new ventures successfully. The CDO structure was in place, and would soon prove an effective vehicle for commercializing two break-through plastics discoveries.[11]

Tragedy struck the division in 1956 when escaping solvent vapors in the Schenectady alkyd resin plant exploded, killing three employees and substantially destroying the facility. The new investment costs of rebuilding the alkyd resin business were daunting, and no amount of blue-sky optimism could forecast that the business would again earn a satisfactory return on the investment. GE therefore sold some alkyd resin assets to Archer-Daniels-Midland, and later sold the Anaheim, California, satellite plant to Chevron.

A similar accident occurred the following year in the Pittsfield phenolics plant, also with three fatalities. This time Gibson immediately determined to stay in the business, ignoring demands of GE transformer executives to get his explosive stuff out of their Works, and he took emergency steps to supply GE phenolics customers with product, which GE bought from competitors. He also hired a professional safety engineer from outside GE to improve safety emphasis and training throughout the division.

Company confidence in Gibson's leadership caused the GE Conduit Products, Wire and Cable, and Wiring Devices Departments to be added to the division in late 1958. With this change Gibson moved his headquarters from Pittsfield to Bridgeport, Connecticut. There was still no thought at the top of the company that chemical-type products would become a major factor in General Electric growth. *Time* magazine's January 12, 1959 issue featured CEO Cordiner on the cover, but the story focused on nuclear power and did not mention plastics or chemicals.

General Electric managers outside the chemical business were surprised by the 1959 announcement that the molded-plastics business, including the Taunton, Massachusetts, and Decatur, Illinois, plants had been sold to Haveg Industries. This disposition was a near first in GE corporate experience, but it reflected volume stagnation and marginal profitability of what had once been the largest product line of the division. Custom molding had become a business in which the division management did not choose to compete further.

In 1960 Gibson returned to Pittsfield as vice-president in charge of the much larger GE Transformer Division, and Reed succeeded him as Vice-President of the Chemical and Metallurgical Division. Also of great future significance in the mid-1950s, Abraham L. Marshall's scientists at the GE Research Laboratory made three important discoveries:

1. The announced synthesis of small diamonds, the first in the world;

2. A remarkable thermoplastic polycarbonate polymer;

3. A polyphenylene oxide polymer formed from xylenol by a novel oxidative coupling reaction.

CHAPTER 4

Loctite

An Invention That Got Away

While the new Chemical Division and the GE silicones venture were getting underway (1945–1952), Research Laboratory chemists Birger W. Nordlander and Robert E. Burnett continued a program to create novel electrical insulating varnishes. In most of these Permafil varnishes a special solvent polymerized with the resin during cure, yielding a solid impregnant structure after baking that was more free of voids than usual. This was desirable from an electrical performance point of view.

Nordlander named one unusual material from this program "Anaerobic Permafil," because it rapidly polymerized at room temperature in the absence of air. For storage, the product required a continuous air-bubbler to maintain its liquid state. Nordlander and Burnett prepared Anaerobic Permafil (see Figure 4-1) by oxygenating tetraethylene glycol dimethacrylate to an unstable epoxide, which remained a thin liquid as long as it was kept in good air contact.

GE's Chemical Materials Department offered the unusual technology to customers as the monomer, along with instructions on how to oxygenate it for use as a sealant or glue. Not being able to offer a packaged, ready-to-use material was a very awkward marketing approach, though several customers did use it for sealing metal joints or cracks. In 1952 GE dropped the material from its product line and arranged for another supplier to accommodate the few customers. As no patent license was negotiated, the technology

45

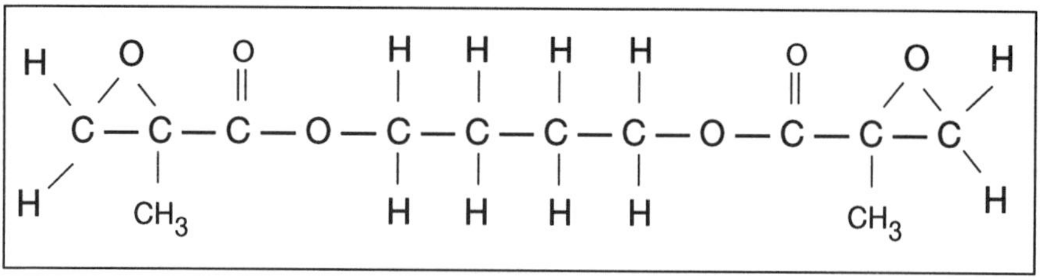

Fig. 4-1. Anaerobic Permafil

effectively went into the public domain.

Vernon K. Krieble, head of the chemistry department at Trinity College, became interested in the technology through his son, Robert H. Krieble, then a manager with GE's Chemical Division in Pittsfield. R. H. Krieble had joined the GE Research Laboratory in 1943, worked there in silicone research, and later transferred to the Chemical Division in Pittsfield, becoming R&D manager of the Chemical Materials Department and later general manager of the Chemical Development Operation (CDO). CDO at that time was developing a polycarbonate polymer, which is described in Chapter 6. Krieble left GE in 1956 to join his father in commercializing a new sealants product that stemmed from the Nordlander/Burnett discovery. Bob Krieble's GE acquaintances were surprised he would forgo a promising GE career for a risky start-up venture with an unknown specialty adhesive.

During Loctite (originally American Sealants) Corporation's early years, few details were generally known about the privately held company, which was headquartered in Newington, Connecticut. How this extraordinarily successful chemical specialties company got started is told in R. H. Krieble's 1980 account.[1]

The anaerobic adhesive story starts with the discovery in the General Electric company Research Laboratory by Nordlander and Burnett of a substance they called "Anaerobic Permafil." It was comprised of a dimethacrylate monomer carefully oxygenated at elevated temperatures and then stored with a stream of air bubbling through it. A drop of the material placed between crossed microscope slides or metal strips rapidly polymerized at room temperature, thereby bonding the two parts together. They applied the word "anaerobic" to this behavior, indicating that it came to life in the absence of air.

Although the material failed to achieve general use, Nordlander and Burnett had discovered a material with extraordinary properties. I happened to describe the situation to my father, (Vernon K. Krieble) who was then head of the chemistry department at Trinity College. He was intrigued and said immediately that the answer to the problem of the air hose was to increase the material's sensitivity to oxygen by one-thousand fold. One of the things he tried was to oxygenate a similar molecule but without the polymerizable double bond, namely tetraethylene glycol dimethyl ether.

Sure enough, when this was mixed with the monomer, he obtained an anaerobic cure, and discovered that the mixture could be stored for years in polyethylene bottles, which are permeable to oxygen. He had indeed achieved that thousand-fold increase in oxygen sensitivity."

With a satisfactorily packaged product now in hand, the issue became whether the novel sealant would find enough uses to make a commercial venture successful. Partners in the new company optimized the product to improve its adhesion and developed a test kit of varying-strength adhesives and a primer. The idea of using Anaerobic Permafil to lock nuts and bolts together had been disclosed in the Nordlander patent. A handful of Connecticut companies tested the product as a locking glue to hold nuts and bolts secure against loosening under vibration. These customers were pleased with its performance, and so Loctite management targeted thread-locking as a major application. The Kriebles commissioned a Batelle Memorial Institute study, which reported a large market for thread-locking devices, one of which was the familiar lock-washer under a tightened nut. Battelle suggested pricing the glue so that its use by customers would cost somewhat less than the equivalent lock-washer cost. R. H. Krieble says that Loctite managers started with this benchmark for initially pricing a 10-cc bottle at $4.50. (Assuming a quantity discount of 50%, and subtracting an estimated packaging cost of $1.00/unit, the material price equates to more than $50 per lb, inferring that a large margin could be available to pay for effective distribution.)

Early on, when the company could only afford two sales representatives, it relied on broad publicity to spread the word through the mechanical-design industries, and to generate orders (not free samples) for the test kits. Loctite then quickly established nationwide sales coverage through manufacturers representatives, using an attractive incentive commission schedule. The product line was expanded to five different locking strengths, a new design concept, for which Loctite developed extensive performance data. Inquiries reached 1000 per week. Loctite grew rapidly and was profitable from the very beginning. With U.S. sales growth well underway, Loctite also established early distribution overseas. Success in this effort eventually built a volume nearly equal to that in the U.S.

Eastman Chemical in the 1960s announced a competitive technology: cyanoacrylate adhesive. Loctite took a license for this product and broadened its line with this fast-curing material. Later acquisitions of Permatex and Woodhill Chemical further expanded the Loctite sealants and adhesives line, and its distribution channels.[1,2]

Perhaps the most remarkable aspect of Loctite Corporation's growth was the self-financing. Krieble says a 1956 loan of $100,000, for debentures convertible to 25% of the common stock, was the only venture capital required in addition to the founders' own backing. A simple manufacturing process, high margins from the value pricing decision, and rapid product acceptance allowed the business to finance its growth through retained earnings, and later from borrowing on the credit of the corporation. The privately held Loctite Corporation made a first public stock offering in 1981, and sold 30% of the shares to the German Henkel Company. By 1995, Loctite sales had grown to $785 million and net earnings were $84 million. Interestingly, this sales volume was roughly comparable to that for GE Silicones. Henkel acquired the 65% remaining Loctite shares in 1996 with a bid of $61 per share, indicating a market value for the company of around $2 billion.[3] One may infer that the Kriebles joined the ranks of decimillionaires.

Lessons Learned

From GE's point of view in hindsight, this was an invention that got away. GE Chemical Division managers failed to envision the magnitude of commercial uses for the Nordlander/Burnett technology that might be developed if the air-bubbling storage requirement could be eliminated. Nor is it recorded that any GE chemist suggested how the product could be modified so as to use air-permeable polyethylene packaging. GE developed a successful electrical varnish line from the Permafil technology, but achieved an annual sales volume of no more than $5 million. Loctite's history showed that by ingenious modification of the technology, plus aggressive marketing, a solid product beginning for an eventual $1 billion sales corporation could be generated. Over time the extraordinary financial returns possible from successfully marketing a truly innovative chemical specialty worldwide were demonstrated. R. H. Krieble states that the company earned 30% return on equity over its first two decades. Loctite Corporation's 1990s sales and earnings were roughly comparable to those of GE Silicones, but Loctite's economic performance from start-up to maturity was far superior to GE's return in silicones with respect to return on investment and cash flow.

CHAPTER 5

Synthetic Diamond

GE Breakthrough Caps
Two Centuries of Research

In fiction, creative quests are carried out by romantic, intuitive loners. In reality, nuclear submarines, genetic engineering, trips to the moon, and the synthesis of diamond were achieved by teams of people with their feet firmly grounded in science and engineering. The drama of nearly two centuries of diamond-making attempts culminated in GE's announcement of the first reproducible process for diamond synthesis in February 1955. In that breakthrough, intrigue and intuition played smaller roles than thermodynamics and teamwork. The invention also awaited the investment of large (though not enormous, by modern standards) resources of people and money. Corporate research laboratories could apply these resources with the requisite patience. Diamond making was thus the fruit of a style of science, a mode of organized invention, and a type of institution that all ripened in the mid-twentieth century.

The Early Quest for Diamond

The earliest observations that diamond burned to carbon dioxide, implying diamond to be pure carbon, are often credited to Smithson Tennant in 1797, and later scientists confirmed the chemical equivalence of the two carbon forms. By 1813, when Humphry Davy set diamonds on fire with a burning lens at the Duke of Tuscany's palace near Florence, the feat was a known scientific stunt. Davy's protégé, Michael Faraday, witnessed this demonstration and soon repeated it himself. Ever since the equivalence proof, a lengthening line of unsuccessful experimenters, including Faraday, searched for clues on how to convert graphite to diamond.

One intriguing effort came in 1880, when James B. Hannay tried to make diamond by heating mixtures of hydrocarbons, "bone oil," and lithium to red heat in sealed metal tubes. Most of his tubes exploded, but three held. Inside them, he found tiny hard crystals that he believed to be diamond.[1] In 1943, X-ray crystallography indicated that the Hannay fragments, then in the British Museum, were not only diamond, but diamond of a type rarely found in nature.[2] But analyses in the 1950s reversed the verdict. In 1962, Kathleen Lonsdale wrote "a near certainty that these Hannay specimens are natural and not synthetic diamond."[3]

Henri Moissan, an early Nobel laureate in chemistry, believed that he had made diamonds in the 1890s by a process of dissolving sugar charcoal in molten iron, and quenching the solution in cold water. The supposedly great pressures produced by the contracting, solidifying mass of iron would, Moissan believed, convert the charcoal to diamond. Transparent, highly refractive carbon-containing crystals indeed resulted.[4] No tests conclusively indicated them to be diamonds, however, and attempts to replicate Moissan's work produced only silicon carbide fragments.

Charles A. Parsons, an inventor of the modern steam turbine, made a systematic attempt to replicate the work of predecessors in the 1920s, and added a number of other experiments of his own, involving high-pressure apparatus or projectiles fired into tapering cavities in steel blocks to achieve simultaneous high pressure and temperature. He too thought early in his efforts that he had made diamond.[5] But he scrupulously reinvestigated and found that his processes had produced other crystals. A 1928 survey recorded Parson's conclusions that neither he, nor anyone else, had succeeded in making diamond.[6] Despite other claims of success, one recent survey states, "It is sufficient to say that there was no clear, unequivocal evidence to support any of the claims of success in synthesizing diamonds."[7]

Thermodynamics and High-Pressure Science

Modern science supplies the search for diamond with a map and an exploration vessel. The map was provided by the application of thermodynamics to chemistry. The vessel grew out of pioneering work on sustaining high pressures in laboratory apparatus that would win Harvard professor Percy W. Bridgman a Nobel Prize for physics.

Diamond's higher density than graphite, and the location of rich diamond mines near magma "pipes" penetrating deep into the earth, had by 1900 led to the supposition that diamond can be formed by putting graphite under high pressure. Thermodynamics explained why. In the late nineteenth century the work of Rudolf Clausius, J. Willard Gibbs, Svante Arrhenius, Jacobus H. van t'Hoff, Wilhelm Ostwald, and others introduced the pressure–temperature phase diagram and the idea that in each region of the diagram, the preferred phase of the material was the one with the lowest free energy. Thermodynamics offered ways to calculate free energy as a function of pressure and temperature. The additional theory of chemical kinetics explained why materials often persist for very long times as interlopers in thermodynamically unfavored regions. For example, no one has yet observed a diamond ring revert spontaneously to graphite, even though graphite is the lower free-energy form at room temperature and pressure.

Theorists also speculated that if high free-energy, chemically combined carbon could be released in a metastable, or transient state by some energy transfer, the carbon might recombine in the diamond state without resorting to high pressure.

Meanwhile, experimenters and theorists were filling in the pressure–temperature map. In the 1930s Frederick D. Rossini and Ralph S. Jessup measured the combustion of diamond and graphite to carbon dioxide in careful experiments. They calculated the phase diagram equilibrium line between the carbon-stable and diamond-stable regions from 13 kilobars at low temperatures to 39 kilobars at 1000°C.[8]

But Bridgman's work in what should have been a diamond-stable region did not convert graphite to diamond at what he measured as 70 kilobars and 600°C[9] (although later corrections suggest the pressure was more like 50 kilobars).

In a 1939 Russian publication little noticed in the West, Leipunskii had also presented a linear extrapolation of the equilibrium curve, and he concluded that pressures of about 60 kilobars and temperatures of around 1700°C would be needed to directly transform graphite into diamond. He also speculated about catalytic transformations. With the use of such solvents as iron, he proposed that pressures around 40 kilobars and temperatures of 600°C. would be sufficient.[7,10] This was not too far outside the boundary already established by Bridgman's work. Leipunskii had experimented with mixtures of molten metals and silicates saturated with carbon, but at relatively low pressures. Bridgman notes that when he first achieved high pressures in his apparatus, "graphite was tried, for obvious reasons." No diamond resulted.[11]

In the 1930s, Bridgman's equipment had reached regions where steel became as yielding as rubber. He contacted metallurgist Zay Jeffries, who was leading the GE effort to introduce cemented tungsten carbide, a hard cutting-tool material, into the United States. By 1940 Jeffries was supplying Bridgman with cemented tungsten carbide parts and Bridgman was opening up new pressure realms.[12] It was known that at atmospheric pressure diamond-to-graphite conversion proceeded spontaneously at 1500°C, suggesting that the reverse transition might occur near this temperature if the pressure could be raised high enough.

Industrial diamond was needed in World War II to grind tungsten carbide for cutting tools and armor-piercing shells, but it was in short supply. In 1941 three companies, GE, Carborundum, and Norton, sponsored an effort by Bridgman to make diamond. Adopting the high-temperature, high-pressure approach, Bridgman explored transformations in both directions, graphite-to-diamond and diamond-to-graphite. His experiments ranged from applying 400 kilobars of pressure to graphite at room temperature, to more than 40 kilobars at red heat. But he did not make diamond.[13]

The Synthesis of Diamond

The wartime project ended and Bridgman returned to other research, but the motivation to synthesize diamond remained. The material was industrially important and expensive, and scientists had good reason to believe it could be made in the laboratory. Among those that acted on that belief after World War II were ASEA in Sweden, and Union Carbide, Norton, and GE in the United States.

ASEA had begun work on diamond synthesis in 1942. Its program built on inventions and the leadership of Balthazar von Platen, who had been a pioneer in the (unrelated) field of refrigeration. His was an extraordinarily complex apparatus, a veritable Rube Goldberg device. An outside cylinder containing water could be pressurized to 6 kilobars. Inside was placed a watertight copper sphere "pressure intensifier," with six internal pistons arranged spherically to bear on six sides of a sample cube. The carbon sample, encased in thermite powder, was enclosed in the copper sample cube. Heat was provided by thermite combustion ignited electrically. After interruption during the war, the ASEA project resumed and continued into the 1950s under Erik Lindblad.

According to John C. Angus (1994),[14] W. G. Eversole, working at the Linde Division of Union Carbide, succeeded in 1952 at growing diamond from carbon monoxide onto diamond seed crystals at temperatures of about 1000°C. and pressures of only about 0.1 kilobar. Although Eversole was later granted two patents (1961), he did not otherwise publish his results, and the metastable route remained an intriguing possibility rather than a proven technique. Documented metastable diamond making was achieved only long after the commercial development of high-pressure processes.[15]

The Norton Company had taken over the apparatus used in Bridgman's wartime project. There Loring Coes, Jr., and Samuel Kistler carried on the work and believed that they had produced traces of diamond. However, Norton abandoned the diamond-making effort, despite having successfully pioneered the high-pressure synthesis of several semiprecious minerals: garnet, zircon, tourmaline, beryl, topaz, and an unusual new dense form of silica. In October 1950, Norton contacted GE, a cosponsor of the wartime investigation, and proposed a joint project.

C. Guy Suits, GE Vice-President and Director of Research, initiated a literature review and, with several of the Research Laboratory scientists, concluded that diamond synthesis might well be feasible, either by high-pressure reaction or by some metastable reaction at low pressure. At Suits' initiative, GE decided in the spring of 1951 that, rather than participate in a joint project, it would undertake its own "investigation in the area of high-pressure synthesis with diamond as the primary objective." GE declined the Norton offer and took care not to receive any proprietary information.[16,17]

The GE program also included a parallel effort looking for possible short-cuts to diamond. A research program headed by Richard E. Oriani and David Turnbull investigated metastable processes at low pressure. It was unsuccessful, and was discontinued in the mid-1950s.[18]

The larger effort, a group led by Anthony J. Nerad within Abraham Marshall's organization, grew to include Francis P. Bundy, H. Tracy Hall, Herbert M. Strong, Robert H. Wentorf, Harold P. Bovenkerk, and James E. Cheney. They explored high-pressure, high-temperature syntheses. In 1951–1952 Bundy began experiments using a 400-ton press and steel anvils developed from Bridgman's designs. The team initially used them to explore the equilibrium line between graphite and diamond. An early contribution, by Louis Navias, was the use of the common natural stone, pyrophyllite, as a ceramic sealing material in place of the more difficult to obtain pipestone used by Bridgman.

Through 1952, Bundy and Strong explored improvements in the pressure vessel, while Hall and Wentorf looked at chemical reactions that might make carbon available for diamond formation. By early 1953, GE had placed its first firm points on the pressure–temperature map. It also confirmed that, in Strong's words, "activation energy for the transition of diamond to graphite increases with pressure and slows down the conversion."[19] This appeared to be the possible "show-stopper" that might put diamond perpetually out of reach. It made more urgent the need to try catalysts that would

Fig. 5-1. GE Research Laboratory diamond research team: from left, F. P. Bundy, H. M. Strong, H. T. Hall, R. H. Wentorf, A. J. Nerad, J. E. Cheney. ca. 1955

reduce the activation energy and thus speed the conversion.

The GE team also recognized that for an eventual commercial process a single-stage pressure vessel would allow more frequent runs and result in lower production costs than a multistage apparatus. Pursuing better single-stage methods, Strong and Hall independently explored designs involving a piston pressing into a cylinder. Then, to double the stroke of the piston, Hall proposed to use two of the conical pistons, pushing from top and bottom into a chamber supported around the middle by, in his words, "a belt of exceedingly strong metal backed up by binding rings."[20] The "belt" was born. It became the principal vessel for bringing back diamonds from the realm of superpressure.

Through 1953, the team explored a variety of approaches, ranging from experiments with new pressure vessels to attempts at metastable synthesis. Occasional glinting crystals and faint diffraction patterns gave a tantalizing "smell" of diamond, but no more. In early 1954, Hall proposed making the inner cylinders of his belt out of cemented tungsten carbide rather than steel. This, in retrospect, was perhaps the improvement that brought the apparatus into the pressure region where diamonds could be made.

However, the GE team remained poised tantalizingly on the brink of diamond for nearly a full year before achieving success. Experiment after experiment proved either negative or inconclusive. GE research management's patience threatened to wear thin. Bovenkerk and Strong (personal communications to Wise) recall that in late 1954 they were told by Nerad that the money and time limits of the project were about to be reached. The scientists urgently wanted to continue and arranged some "bootleg time" to circumvent the project-accounting tally.

Meanwhile, the team had bought some natural diamond fragments, and sought to grow more diamond on them. In December 1954, Strong wrapped one such crystal in iron foil, immersed it in a commercial compound that is used to put carbon in metals, and sought to subject the combination to 54 kilobars and 1200°C. The experiment, labeled "Run 151," was carried out (again, at substantially lower pressure than the experimenters believed) and the results went to metallography. A week later the report came back. In addition to the expected seed diamond, the Run 151 sample contained two other fragments, "one shaped like a brick out of an old brick wall with some mortar clinging to it, and the other like a bluntly pointed and slightly flattened rod." A series of tests, culminating in the decisive X-ray diffraction test, revealed them to be diamond.

This unexpected result provided the impetus for Hall to put iron catalysts along with the graphite placed in his belt apparatus. He was able to reach 75 kilobars (he thought it was 100 kilobars) and 1200°C. and keep the sample at pressure for a few minutes during a 20-minute cycle. On a tantalum disk at the top of his apparatus emerged, in his words at the time, "dozens of hard (makes terrific scratches in glass) transparent, isotropic (under Navias' petrographic microscope) triangular faced crystals." X-ray diffraction tests soon confirmed that these crystals too were diamond.[20]

In the weeks that followed, Hall's work proved repeatable, but the Run 151 result did not. It took nearly forty years to clear up the mystery of those initial diamonds. Tests in 1993 on the only surviving Run 151 diamond showed it to be a piece of natural diamond. Apparently it had been inadvertently included in the experiment. GE published an explanation modifying the historical record.

Back in January 1955, subsequent experiments with Hall's belt showed that the process could be replicated, and that other metals than iron in Group VIII of the periodic table would catalyze the production of diamond at pressures above 50 kilobars and temperatures above 1200°C. Those conditions penetrated into both the diamond stable region for carbon, and the liquid region for mixtures of carbon and a Group VIII metal. The mechanism apparently involved dissolving carbon in the molten metal, followed by its re-crystallization as diamond.

GE announced these results in February 1955, and a paper in *Nature*, by Bundy, Hall, Strong, and Wentorf, also marked the culmination of the scientific quest for a reproducible diamond-making process.[21] Many highlights remained ahead, such as Wentorf's stunt of converting peanut butter to diamond, direct synthesis of diamond from graphite without catalyst by Bundy in 1963,[22] synthesis of gem diamond by Strong and Wentorf in 1970,[23] and the much later development of low-pressure processes by many investigators. But GE's project and Hall had made the crucial breakthrough. Hall was disappointed that the GE announcement did not emphasize his singular contribution. He left GE, accepted a leadership position at Brigham Young University, and there continued to advance high-pressure research.

Some years after the GE announcement, ASEA claimed to have had success in converting some graphite into diamond in February 1953. Inexplicably, they had made no announcement at the time and never offered a technical publication. Why did the ASEA research group not announce their achievment at once? According to Halvard Liander (1980), a leader of the effort:[10] "The answer is quite simple. We did not wish to draw attention to our work until more information had been gained. The von Platen equipment was very complicated and cumbersome to work with. A new design of high-pressure apparatus was being planned with a smaller reaction volume, but allowing one or two experimental runs daily instead of monthly, at best, with the original apparatus. And there seemed to be no competitive work going on."

The reader seeking more detail about the history of diamondmaking attempts may wish to read R. M. Hazen, *The New Alchemists*.[24]

The Discovery Heard Round the World

The GE news conference at the R&D Center in February 1955, announcing the project team's success and (in general terms) Hall's process for synthesizing small diamond fragments, created waves in many directions. A *Wall Street Journal* reporter asked Chief Executive Ralph Cordiner if the diamond announcement should affect the price of GE stock. Cordiner thought not, but the reporter replied, "I've just called my office and you're up four and a quarter points." Incidentally, De Beers Consolidated went down two points on the London Exchange. Both "recovered" the next day.[25] Scientists everywhere were impressed. Customers worldwide for mesh-size natural diamond, called "crushing boart," were eager to learn if GE could scale up the process successfully to offer a synthetic product.

The market for industrial diamond in 1955 was well-defined in the United States as the manufacturers of resin-bonded diamond grinding wheels. Diamond is the hardest abrasive known, and the main use of the diamond wheels was to sharpen tungsten carbide cutting tools. The world market had been supplied with natural "crushing boart" for many years as a byproduct of the gem diamond business monopoly of De Beers Consolidated Ltd. Diamond saws were not a large market factor in 1955.

Profit-and-Loss Test in a Hurry

Chemical and Metallurgical Division manager Robert L. Gibson lost no time confirming with Cordiner the assignment of this potential product to a newlyformed product section within the Metallurgical Products Department of his division. (Another GE organization seeking the diamond assignment was the Schenectady Carbon Products section, which made brushes for electrical machinery, and which had furnished various carbon samples to the researchers.) The assignment to Metallurgical Products proved an excellent fit, because large, specially formed tungsten carbide "dies" and "anvils" were key parts of the diamond process apparatus, and the department was itself a major customer for diamond grinding wheels. The Detroit-based business was the largest department in the division at that time and could readily furnish people and space to get on with replicating and then scaling up the Research Lab results. The R&D Center scientists would continue high-pressure research in Schenectady, but a rapid transition to a profit-and-loss department would test the technology and put the product before customers in the shortest possible time.

Almost immediately potential customers such as United States manufacturers Norton and Carborundum were in touch seeking samples. Strategic materials planners for the Defense Department were also interested because natural industrial diamond had been critically short during World War II and again during the Korean conflict. As a consequence, the Defense Department imposed a secrecy order to prevent additional technical disclosure. This created an extreme problem for GE, as some interpretations of the order would prevent GE scientists from talking to others in the company. After satisfactorily clearing up the internal communication requirements, GE's problem with the secrecy order became prevention of the time-urgent right to file for patents outside the United States. Not until 1959 did GE succeed in having the secrecy order lifted so as to continue with important foreign patent filings.

Scale-Up

Technical papers by GE authors and various GE patents, which were first published and applied for around 1960, reveal that small diamond crystals had been produced by subjecting carbon dissolved in a molten metal catalyst from Group VIII of the periodic table to high pressures (above 50 kilobars) and temperatures (above 1200°C). Pressure was applied by a vertical hydraulic press capable of several hundred tons force, and was exerted by "anvils" on the top and bottom of the ingredients "cell". The cell ingredients surrounded by pyrophillite stone were contained horizontally by a circular "die" of rigid tungsten carbide, called a "belt," itself constrained by a steel cylinder. Electric current from electrodes at the top and bottom of the chamber heated it to the desired temperature. Pressure and temperature were maintained for the time period needed to melt the metal catalyst, dissolve the carbon, and grow the desired diamond.[26–32]

As with many revolutionary developments, the process was difficult to reproduce and results were variable. The first diamond product section manager, J. Stokes Gillespie, visited the Research Lab and listened to conflicting technical reports. He then drove to Worcester, Massachusetts with a small vial of the new diamond to inform the Norton Company of GE's success. Microscopic examination by a Norton diamond expert yielded the opinion that these black fragments, so different from natural diamond boart, could hardly be useful in a Norton product. This disappointing feedback stimulated the Schenectady scientists to make more diamond and to fabricate some tiny resin-bonded grinding tools. Similar tools were prepared in Detroit, and the tests by grinding tungsten carbide seemed to go well on this tiny scale. Norton's own tests with additional product were then successful in resin-bonded wheels; and Gillespie set in motion high-pressure apparatus purchases in Detroit.

William K. Cordier directed the first pilot plant and semi-works-production press installations in empty sheds that had been used during World War II to fabricate tungsten carbide armor-piercing projectile noses. GE also set up an application development laboratory where diamond wheels could be made and their performance tested, just as in tests by customers. From time to time in the early years GE managers would debate whether to integrate forward by making and selling wheels as well as the synthetic diamond, but always the decision was made to remain at the industrial-diamond-grit product level.

The early research process produced a tiny diamond yield, a small fraction of one carat per run. And the expensive tungsten carbide dies and anvils tooling would often break after a few runs. Massive improvement was needed in each of these factors before synthetic diamond could possibly become a commercial success. To develop the process and facilitate communication with R&D Center scientists, Bovenkerk and Cheney, each a member of the original Research Laboratory project team, joined the commercial project in Detroit, and each made important technical contributions over long and productive careers with the GE diamond business. They achieved dramatic process improvements over two to three years in the commercial development atmosphere. Working with vendors, new hydraulic press types were

Fig. 5-2. GE Man-Made® diamond in saw blades, wire saws, and core drills, is used to quarry and shape natural stone and masonry, and to renovate concrete highways, bridges, dams, and power plants.

developed and press sizes were greatly increased. Product quality improved, and diamond yields per run increased by many orders of magnitude. Carbide tooling life increased manyfold.

By 1957 GE offered a commercial diamond product for making grinding wheels, priced at a premium to the natural diamond competition. The diamond grit desired by wheel manufacturers ranged in size from 325/400 mesh up to 80/100 mesh. De Beers had sold "bags" of assorted industrial natural crushed boart at fixed prices on pretty much a "take it or leave it" basis, similar to their commercial practice of offering nonnegotiable "sights" of uncut gem diamonds. GE's improved process proved capable of growing diamond crystal to the size range desired, but even after special surface cleaning the crystals were nearly black, in contrast to the off-white cast of the natural crushed grit. The synthetic crystals also had higher aspect ratio (of length to width), and were more friable than the natural crystals.

But from the very beginning, tests of *resin-bonded* wheels made from GE diamond in their application laboratory, as well as those made by the Norton and Carborundum companies, showed superior grinding effectiveness on tungsten carbide compared to the natural diamond counterpart, sometimes by a factor of 2. Fortuitously, GE's original process yielded more friable crystals than the natural and this characteristic was key to superiority in this particular application.   On the other hand, this type of synthetic diamond was unsatisfactory in *metal-bonded* wheels or saws.   These findings clearly showed that different types of diamond crystal would be optimum for different uses.

The "Man-Made" (GE trade name) diamond superiority in resin-bonded wheels was further increased when Norton developed a new bond "B57" wheel, optimized for the new diamond. The GE development program also kept improving crystal quality for this first use.   Norton showed its confidence by placing an early 100,000-carat order (ca. $600,000 at that time) with GE.   This customer's enthusiasm plus other broad demand for the GE diamond led to capacity expansion with many more presses in a new building behind the Detroit tungsten carbide plant.

Instead of the usual practice for forwarding an appropriation request via channels to the GE Board of Directors, Gibson arranged for department manager, W. Kenneth Beardsley, and then diamond section manager, John D. Kennedy, to appear at a board meeting in New York. There they showed some quart bottles of the product, each bottle worth about $50,000 at $6 per carat (5 carats = 1 gram).   This caused great excitement and prompt approval of the expansion.   The diamond product section turned a profit in 1958, just four years after the research discovery.

The GE diamond management team set up Unites States distribution by direct sales contacts with the largest customers.   To reach the large number of small-wheel manufacturers, the Van Itallie Company, an established distributor of natural industrial diamond, was appointed to handle the GE line. De Beers reacted by cutting off Van Itallie as a natural-industrial-diamond distributor.   Van Itallie initiated the service of separating GE bulk diamond by crystal size, and selling different size ranges separately.   GE later adopted this same strategy for its whole line.

International distribution of GE diamond moved ahead rapidly, as the major diamond wheel-makers were calling from abroad to connect with this new source.   The Winter Company, an important German diamond tool maker, became an early GE customer, as did Asahi Industrial Diamond Company in Japan. The commercial practices of De Beers while a single-source monopoly for natural industrial diamond had been difficult for customers over many years, so the GE competitive entry was very welcome. Overseas orders were initially handled by the chemical export sales unit of International General Electric Company based in New York, but were later transacted directly with the Detroit diamond section.   GE contacted European customers by a combination of direct-sales representatives and distributors, while in Japan, the Mitsui Company became an effective sole distributor. By 1960 GE's diamond for resin-bonded grinding wheels (trade-named RVG) was well established around the world.

A consumer-product application for selected crystals as phonograph pickup needles in GE stereo equipment made good publicity, but provided trivial volume.

Competitive Challenge

De Beers competed hard against invasion of its market by the synthetic product. In 1962 their expenses-paid seminar in Paris for worldwide industrial diamond customers touted their "selected natural diamond" (SND) as superior to the GE product. But four years later the claim of a similar seminar held in Oxford was "De Beers synthetic is better than GE synthetic." De Beers had made a major commitment to synthetic-industrial-diamond research immediately following GE's 1955 announcement.

It became evident from trade reports and finally in the marketplace that De Beers had organized a top-rank technical team in South Africa, to develop for themselves the diamond-making technology now known to be feasible from GE's announcement, publications, and commercial product. After four years of effort De Beers succeeded in producing synthetic diamond, and in 1961 they began offering product from a factory at Springs, South Africa. Some years later they established a second factory in Ireland. General Electric by this time had issued patents in some countries, and five in the United States. When the United States secrecy order was lifted in 1959 GE immediately filed in other countries, barely gaining time priority over De Beers in South Africa. The delayed GE filings were opposed by De Beers, and GE was similarly opposing those filed by their competitor. South Africa patent law relies on *opposition* rather than on *examination,* as in the United States.

Decisions regarding GE patent filing and licensing policy abroad had long been the responsibility of a licensing group within the International General Electric Company (IGE), but fundamental policy differences promptly surfaced between IGE and the diamond business managers, the latter believing that the advantages gained by a patent position should be preserved if at all possible, and that any licensing should reflect the high value of the technology. Charles E. Reed, who had succeeded Gibson, and Cordier, who became manager of the diamond section in 1962, prevailed in this corporate argument, and the company made a clear assignment of international diamond patent responsibility to the Detroit business management. The decision to seek and enforce patent rights by expensive and chancy litigation wherever possible worldwide was not taken lightly, but Cordier continued suits against De Beers in the latter's home country and elsewhere. This was a break from GE tradition, but he was backed in the initiative by division manager Reed. Thus, expensive litigation in South Africa and other countries continued for several years, finally resulting in a clear South Africa judgment in favor of GE. As the court actions proceeded, De Beers sought an accommodation through licensing at a low figure, but finally agreed to a complex proposal at GE's high value estimate. GE also litigated in other countries, successfully against Russian diamond in Germany, and ultimately, with limited success, in Japan.

Diamond for Metal-Bonded Wheels and Saws

Sales of GE RVG diamond for resin-bonded grinding wheels were growing rapidly and prices had been reduced to the $1.00–$1.50 per carat range. But a different diamond crystal was needed for metal-bonded wheels and saws, one with fewer inclusions, less friability, and larger size. Diamond saws are round steel discs with the outer edge made of a metal-bonded layer of larger diamond crystals. Bovenkerk led GE process development toward that objective and became successful in the 1962–1964 period, when GE introduced diamond for both metal-bonded saws (trade named MBS), and a grade for metal-bonded wheels (trade named MBG), in mesh sizes from 60/80 to 30/40. GE MBS diamond proved another commercial breakthrough, as the saws made from it were greatly improved over those made from natural diamond. Initial offering prices for saw diamond crystal were in the $2.80 to $4.00 per carat range.

The broadened product line firmly established GE's leadership in industrial diamond worldwide and by the end of the 1960's about half the sales volume came from customers outside the United States. As the products and application technology became more complex, GE set up technical service centers in the United States, in Germany, and in Japan. In 1972, at the customer's suggestion, GE made a 10% investment in Asahi Industrial Diamond Company, an interest that was sold many years later. The diamond product line was probably the first in General Electric ever to have greater sales to customers outside the United States than to those within. But the De Beers organization continued to follow GE technical developments and to compete with synthetic as well as natural diamond around the world.

A New Plant and New Organization Status

Continuing rapid sales growth for both RVG wheel and MBS saw diamond products, plus the need for additional laboratory space, pressed the plant capacity limits in Detroit, so a new location with existing factory space was purchased in Worthington, Ohio, a suburb of Columbus. Concurrently, Reed proposed, and received approval for in 1967, upgrading the diamond product section to department level. The newly named Specialty Materials Department thus became organizationally independent of the Metallurgical Products Department (MPD), while continuing to purchase from MPD the tungsten carbide anvils and dies needed in the diamond apparatus, and also sharing some facilities and services. The Worthington facility operated as a satellite plant to Detroit until 1968, when general manager Cordier moved the department headquarters to Ohio and the two plant roles were reversed. The remaining presses were moved from Detroit to Worthington in 1972.

The Worthington personnel plan employed salaried technicians as operators, used no time clocks, and emphasized employee involvement and empowerment, similar to division programs that had proved successful at the Mt. Vernon, Indiana plastics plant and that were being tried at Selkirk, New York (Chapters 6 and 7). This proposal prevailed over strong objections from GE corporate employee relations experts. And like the two plastics plants, the Worthington location has remained nonunion.

Borazon

Theoretical analysis and continuing high-pressure, high-temperature research at the R&D Center by Wentorf yielded a new very hard material, cubic boron nitride (formula BN), in 1958.[33] After extensive development in the following years, this new abrasive material (tradenamed Borazon) was put into production in the Worthington plant. It was effective in the grinding wheels for sharpening hardened tool steels. For this application the competitive grit material was low-cost aluminum oxide. After several years of aggressive GE market development, Borazon wheels had demonstrated superior effectiveness, in some applications by 100-fold, but were substantially more expensive. The wheel manufacturers, some of whom made their own aluminum oxide grit, were reluctant to push the more expensive product, but GE continued the market development effort. Borazon wheels eventually became preferred for grinding the hardest of the alloy tool steels, and they represent a significant business worldwide today.

Fig. 5-3. GE Stratapax® polycrystalline drill blanks are used to improve production in the drilling industry.

Gem Diamond

Strong and Wentorf at the R&D Center discovered a process route in 1970 that could yield gem-size diamonds.[23] These synthetic stones, which could be grown up to 5 carats in size, typically had various color casts, in contrast to the preferred blue-white of the finest natural stones. Also, it took a long time to grow such extraordinarily large crystals in an expensive press setup. In any event, neither GE nor its competitors appear to have achieved a commercial gem-size synthetic diamond offering.

Saw Diamond

The R&D program of greatest commercial significance continued to be for metal-bonded saw diamonds. As the MBS crystal size was increased and crystal strength improved, manufacturers and users of improved diamond saws were able to expand vastly the uses for cutting granite, marble, structural concrete shapes, antiskid highway treatments, and airplane runway surfacing. GE synthetic diamond crystals of this special size and quality made possible today's large applications and markets for diamond saws.

Polycrystalline Diamond Compacts

Natural diamond, which had been largely replaced by synthetic in wheels and saws, had other industrial uses, and these became a challenge to GE researchers. In oil drilling, for example, certain special drills were faced with moderate-size diamond crystals. And some wire-drawing dies, particularly those for tungsten lamp wire, were best made from diamond. While it had proved difficult and expensive to grow synthetic diamond crystal to the required size for these applications, the idea of securely bonding small synthetic diamond crystals together in the size and shape needed was intriguing. Hall, now working at Brigham Young University, demonstrated one product he called Megadiamond that had some commercial success.

Wentorf's research discovered a satisfactory process in 1971, subjecting diamond fragments and cobalt to high pressure and high temperature in the same apparatus in which the commercial industrial diamond had been made. The polycrystalline diamond/cobalt mass could be cut to shape by electric-discharge-machining, or, if cobalt had been dissolved away with acid, by laser cutting. In 1973, GE was the first to put successful polycrystalline diamond compacts, or blanks (trade named Compax) on the market. These were used to make wire-drawing dies and for special metal-cutting tools. These diamond cutting tools are important for machining high silicon–aluminum alloys and, in Europe, for shaping resin-filled wood composites. With this bonded diamond product a marketing policy question surfaced: Would the GE diamond compacts for cutting tools be offered exclusively to the sister GE department, Metallurgical Products, or broadly to all cutting-tool manufacturers? Decision: offer the blanks to all. An improved polycrystalline compact version, especially designed for diamond-drill manufacturers and tradenamed Stratapax, was offered in 1976 and became an important commercial success. GE markets diamond compacts today in many shapes and sizes for different applications: flat, round wafers, triangles, squares, among others.

New Overseas Plant

Continuing volume growth in industrial synthetic diamond, diamond compacts, and Borazon led GE to consider a second plant site. While it would have been feasible to expand only in Ohio, Louis Kapernaros, who succeeded Cordier as department manager, believed that customers worldwide would feel more secure if their main diamond supplier had two producing locations. He and his management team elected to locate within the borders of the European Common Market, so a site near Dublin, Ireland, was chosen and a substantial plant built in 1981. The Ohio plant was also expanded at this time.

Diamond from Explosive Shock

Diamond traces in meteorites and in debris from meteorite impacts led scientists to experiment with explosion shock on graphite. In 1959 Paul S. De Carli and John C. Jamieson at the Stanford Research Institute demonstrated that an explosion shock-wave impact could transform graphite into diamond. DuPont scientists then explored explosion techniques and developed a reliable process that yielded microscopic grain-size diamond ideally suited for the important function of polishing gem stones. Tradenamed Mypolex, this diamond dust sells at a higher price than synthetic diamond abrasives and enjoys a steady market in its special field of use.[24]

Vapor Phase, Low-Pressure Diamond Deposition

Because the high-pressure, high-temperature research and production equipment is expensive, efforts to make diamond at low pressure have continued in many laboratories around the world, and several successes have been reported.[34] They are collectively called CVD, meaning chemical vapor deposition, and are similar to the original findings of Eversole. The carbon source can be carbon monoxide, dioxide, or methane, while thermal activation supplied from a hot filament, radio-frequency (RF) plasma, electric arc, or laser can trigger conversion to diamond. Typically a thin film of very fine diamond crystals is deposited on a substrate. At Pennsylvania State University there is a Diamond and Related Materials Center specializing in the science of such phenomena. A recent (1996) news release from the small QQC Company in Dearborn, Michigan cites a new CVD approach using four powerful and finely tuned laser beams that vaporize some substrate as well as trigger diamond formation and deposition.

Many such announcements predict that the new process will prove to be a more economic approach to making diamond than the high-pressure, high-temperature techniques. Diamond films deposited from these low-pressure processes have proved useful as substrate in certain solid-state electronic components, where the extraordinarily high heat conductivity of diamond is an important use factor. But thus far, a couple of decades of research in vapor phase diamond deposition have not yielded products commercially useful in the abrasive or cutting applications. GE performed extensive research in the vapor-deposition field and established a joint venture with Asahi in 1989 to explore one of the technologies. The venture did not demonstrate important commercial results.

Growth of Competition

As synthetic diamond markets grew, as patents expired, and as the technological knowledge diffused to many places, new competition in industrial diamond sprang up, first with small crystal for resin-bonded grinding wheels, and subsequently with the larger saw diamond crystals. De Beers gradually approached product and sales equivalence with GE. Several Russian producers offered diamond products, as did competition in Rumania. Tomei in Japan became a major producer, while resisting GE efforts to enforce patents in that country. The Iljin Corporation of South Korea established a diamond subsidiary and became a significant producer. Iljin was later proven in a United States court to have purchased, for $1 million, GE technology stolen by a former GE employee who had left the company in 1984. A United States court order for Iljin to cease making diamond for seven years was appealed, and the upshot was an agreement for a licensing fee to GE. A cottage industry of small producers in China now offers small diamond crystal. Prices of $0.17–$.40 cents per carat may be typical for small-wheel crystal, and lower quotations have been reported. Prices for saw diamond may now range from $1.00 to $2.70 per carat for the most common types.

Synthetic diamond now probably accounts for 90% of all industrial-diamond usage, and the combined output of synthetic diamond producers far exceeds the tonnage sales of gem diamonds in the world. The production from GE's Worthington plant has been estimated at 33 tons for the year 1990.

Antitrust Challenge

General Electric, while competing with De Beers and an increasing number of other producers around the world, maintained a low publicity profile with respect to the size and progress of its diamond business. So it was a major shock in April 1992 to be the subject of a United States Justice Department antitrust investigation, with its attendant publicity and serious risks. The action had been triggered by allegations from a former GE management employee that GE and De Beers had conspired to fix diamond prices. GE's own investigation of the charges convinced top management that they were not true, and the employee later recanted with respect to his personal knowledge of such activity. GE declined to plead guilty to a charge it believed to be false, and the case went to court in Columbus in the fall of

1994. Although it was a jury trial, the judge dismissed the action upon the GE motion after the prosecution case had been presented, stating that the evidence presented simply did not support the charges. Some estimates about the diamond industry came to light in the proceedings. The world industrial synthetic-diamond market was said to be about $600 million, of which the combined total of GE and De Beers sales represented approximately 80%. GE's 1993 diamond sales were reported to be $281 million and profits were high.[35]

Summary and Lessons Learned

The brilliant scientific discovery of a diamond-making process, a four-year targeted effort by the R&D Center team, culminated in successful experiments by Hall in late 1954. GE's Chemical and Metallurgical Division promptly established a commercial organization to scale up and test the market possibilities for the synthetic diamond. Despite its disappointing appearance, the new synthetic performed better than natural diamond in resin-bonded grinding wheels, so the new business enjoyed sales demand and market growth as soon as production capability was established. Initial pricing provided good margins from the beginning and early profitability for the product line. GE successfully fought a determined De Beers challenge to the patent position in courts outside the United States and gained an agreement that set substantial royalties as compensation for use of these patents.

GE's continuing product innovations were very successful. Processes were developed to increase yield and to make larger and stronger crystals. These proved ideal for metal-bonded diamond saws. The company then created polycrystalline diamond compacts, used for cutting tools, wire-drawing dies, and oil drills. These two GE diamond developments created an explosion of new uses, beyond anything possible from natural diamond. A different new abrasive, cubic boron nitride (Borazon), created a niche among wheels to grind hard tool steels, while Borazon compacts became useful for some special steel machining.

The diamond business was thus the second new product success of GE's chemical effort, coming soon after the silicones had turned a profit. Lessons from the disappointing silicone experience were well-applied in diamond: no delay in forming a commercial unit with a single-responsibility project manager, rapid development to put an innovative product in front of potential customers, time urgency to establish a world sales and patent position, and continuing product-line innovation and expansion. While diamond sales volume has never been a large share in GE's chemical product totals, the profitability has been an important and consistent contribution.

CHAPTER 6

Lexan Polycarbonate:
1953–1968
The "Unbreakable" Thermoplastic

In 1953, concurrent with the high-pressure diamond synthesis research, Abraham L. Marshall assembled a chemist team at the R&D Center whose objective would be a new wire enamel. The new polymer should have the outstanding strength and toughness of Formex, but must have higher thermal resistance, suitable for continuous service at 130°C. One member of this research team was Daniel W. Fox, the chemist whose discovery of a polycarbonate thermoplastic would trigger GE's entry into the field of engineering plastics.

Fox had come to GE with a different mission. He was born in Middletown, Pennsylvania, spent time in a Bethlehem Steel forge shop, ran a gas station, and in World War II navigated B-29's over Japan. He then came back under the GI Bill to complete a chemistry degree at Lebanon Valley College, earn a Ph.D. in organic chemistry at the University of Oklahoma, and conduct a year of postdoctoral project research there. Fox was hired by GE's Schenectady Works Lab in 1953 to work on electrical insulation for large machinery. As indoctrination to the company he was assigned temporarily to the Research and Development Center (R&D Center) where he joined the wire-enamel project team.

A glyceryl/glycol terephthalate polyester met every thermal, electrical, mechanical and solvent resistance requirement for the wire enamel, but some improvement in hydrolytic stability would be welcomed. Fox recalled a syn-

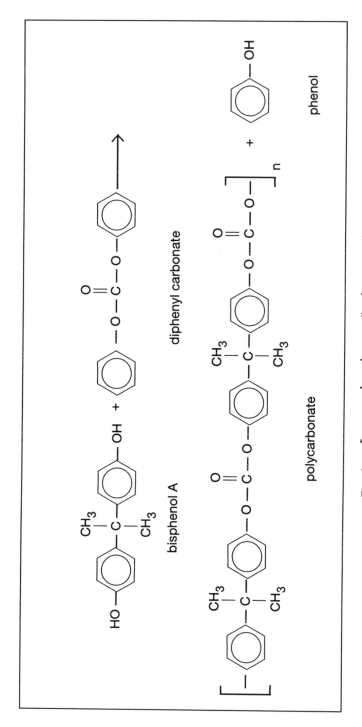

Fig. 6-1. Lexan polycarbonate "melt process."

thesis during his postdoc project, for which he needed guaiacol (o-methoxyphenol). The stockroom had no guaiacol, but plenty of guaicol carbonate. To short-cut a purchasing delay, Fox set about hydrolyzing this carbonate to the phenol in boiling caustic. But the phenyl–carbonate bond would not hydrolyze, an observation Fox says ran counter to what he'd been taught in organic chemistry courses. At the GE R&D Center, Fox wondered if a polycarbonate polymer would show this extreme hydrolytic stability. He tried forming a polycarbonate polymer by transesterification, the process he had been using in the wire-enamel formulations. Reactions of bisphenol A (2,2-bis-(4-hydroxyphenyl)-propane) (BPA) with *dialkyl* carbonates were unsuccessful, but a switch to a melt of BPA and *diphenyl* carbonate produced a polymerization reaction, with phenol distilling off. What emerged after stirring, applying vacuum to remove more phenol and heating to 360°C was a thickening substance that cooled to a solid mass. Fox broke the glass flask off it and was left with what went into GE folklore as "Dan Fox's lollipop," an amber colored chunk on the end of a metal stirring rod. It wasn't just hard, it was unbreakable. You could drive nails with it.

Figure 6-1 shows the reaction, which GE called the "melt process."

The polymer glob had unusual properties, but the Research Lab chemists told Fox that the polycarbonate would have to sit until the Alkanex wire enamel, for which Fox was recognized as coinventor with Frank Precopio, got out the door. He returned to his Schenectady Works Lab position, where he was able to fit in some part-time experimentation with the polycarbonate polymer. His discovery, made within eight months of his joining GE, was a serendipitous byproduct of the wire enamel project. The byproduct would prove far more important than the original objective.[1,2,3]

Polycarbonate Development Transfers to Pittsfield

Alphonse Pechukas, head of the Pittsfield Chemical Development Operation (CDO), visited the R&D Center in 1954, and among other developments was shown the "lollipop." Pechukas had come to GE from Pittsburgh Plate Glass, where his laboratory had worked with *alkyl* polycarbonates, and he recognized that this *aryl* polycarbonate was very different and had some extraordinary polymer properties. Because Marshall and the R&D Center had not picked up on the project, Pechukas started a major polycarbonate polymer program within CDO before transferring to the GE corporate staff engineering group in New York.

Fig. 6-2. Daniel W. Fox, discoverer of GE's first, and most important engineering plastic, Lexan polycarbonate. Photo courtesy of the Othmer Library of the Chemical Heritage Foundation.

The aryl polycarbonate polymer showed impact strength, heat resistance, and dimensional stability far superior to polyamides, such as Du Pont's (nylon) Zytel, the high-performance thermoplastic of that time. The GE team believed the polycarbonate might be developed as an electrical insulation film or as a molding compound, but chose the latter as the first commercial objective. If successful, this would be GE's first *thermoplastic* product, which customers would form by melting and injection molding, in contrast to GE's *thermosetting* phenolics. Thermoplastics melt while being injected into a mold where they cool to a solid. Thermosets polymerize to a solid in a heated mold and cannot be remelted.

The plastics industry at this time was in the early stages of rapid growth. U.S. sales of polyvinyl chloride in 1954 were roughly 400 million pounds, of polystyrene 600 million, and of polyethylene 200 million. A decade later the industry had grown to about 10 billion pounds per year. The higher-performance, higher-price plastics were just beginning to emerge in the mid-1950s. Sales of nylon polyamide molding materials, which had been introduced by Du Pont in 1950, were about 10 million pounds in 1954.

Alternate Polycarbonate Processes

The Dan Fox melt process to make polymers from bisphenol A and diphenyl carbonate could not be scaled up for two reasons: inadequate molecular-weight control and difficulty of handling a viscous melt at high temperature and high vacuum. Further, it was learned that the originally targeted molecular weight was far too high, because it yielded a plastic extraordinarily difficult to mold. So the CDO development team, led by William E. Cass, who had transferred from the Research Lab, and including Eugene P. Goldberg, Kenneth B. Goldblum, E. F. Fiedler, and D. F. Loncrini (and joined by Fox in early 1956), explored a completely different synthesis for the desired polymer. This was an all-solvent system for reacting bisphenol A with phosgene ($COCl_2$). Using methylene chloride solvent, plus pyridine as an HCl acceptor, the team developed a controllable polymerization. A polymer of molecular weight suitable for injection molding could be precipitated after water-washing and phase separation. Recovery and recycling of both methylene chloride and pyridine would be necessary for economic results. The phosgene handling risk was considered manageable, a judgment aided by George E. McCullough's World War II Chemical Warfare training. GE put this process into pilot production so as to sample the market at the earliest moment.[4,5]

A Bombshell from Overseas

After about two years and $2 million of development effort, a pilot plant was operating and the polycarbonate polymer continued to look outstanding. The GE program had tried several variations of the bisphenol A starting material, but none was markedly improved over the original polymer molecule. This was fortuitous because bisphenol A had recently been made commercially available by Dow Chemical, Union Carbide, and Shell for use in making epoxy resins. Fox's patent applications and those of other GE researchers had been filed. Enthusiasm for the program was growing, when in 1956 a startling revelation came from Europe. First, a Belgian patent covering aryl polycarbonates issued to Farbenfabriken Bayer, and then the October issue of *Angewandt Chemie* included a paper by Herman Schnell[6] describing reactions discovered in Bayer laboratories leading to polymers essentially identical with those of the GE development program. It was evident to GE (though not yet to Bayer) that discovery and early research into aryl polycarbonates had occurred independently, and almost simultaneously, in corporate laboratories in Germany and the United States. Publication surely indicated that the German scientists had filed patent applications in other countries. From the Belgian patent docket GE attorneys could tell that Schnell's filing date preceded Fox's first experiment date by a few weeks. So instead of having a world-first polymer invention, GE was certain to be a junior party for patent priority, as well as be facing a formidable research competitor, one of the world's great chemical companies. Declaration and resolution of patent interferences in various countries between Bayer and GE researchers would take several years. The question therefore arose: should GE continue an expensive development program when future adverse patent holdings might put the entire project at risk?

A. Eugene Schubert, who became general manager of the Chemical Development Operation after Robert H. Kreible left GE in 1956 (Chapter 4), had joined the Research Lab in 1943 with a Ph.D. in chemical engineering from Penn State. With Charles E. Reed he worked on methylchlorosilanes distillation and the Waterford silicone plant design. One of Schubert's favorite expressions, "Let's face the facts," fitted the situation. He approached Bayer with the disclosure that GE was also conducting a significant research program in polycarbonates, and he proposed that the winner of any polycarbonate patent interferences license the other on a royalty basis. Such an agreement was negotiated, except that if Bayer prevailed in Europe, they would not agree to license GE there. Both parties were thus enabled to pursue the technology and markets with reduced risk of adverse patent barriers. With a time-equal start the race had begun for polycarbonate technology and market leadership. Schnell's patent applications in the U.S. were eventually granted priority over Fox (they were issued in 1962), meaning that GE would be paying royalties to Bayer for many years.[1,7]

Product and Market Development

Following an employee trade-name selection contest, GE registered "Lexan" for its new polycarbonate and showed the first samples to customers in 1957. Market development manager, William F. Christopher, kicked off industry publicity by personal visits to trade publications. The first Lexan data sheet (March 1957) stated: "Its unusual chemical composition offers a combination of toughness and heat stability not previously available in thermoplastic materials." Limited product availability also required a disclaimer: "Lexan molding compound is not presently available for field evaluation. The present limited quantities of the resin are being used in a controlled testing program designed to provide information needed to arrive at a decision on large-scale commercialization."

Researchers Fox and Goldberg delivered the first GE polycarbonate technical papers at a Gordon Research Conference in the summer of 1957, while a broad sampling program began as material became available. Publicity from frequent speeches by technical and marketing people continued. Christopher formally presented the new material to a packed audience[8] at the January 1958, Annual National Technical Conference of the Society of Plastics Engineers. The most impressive Lexan property cited was the impact strength, >12 ft-lbs per inch, more than double the best nylon materials of that time. The heat distortion point of molded Lexan was at least 100°F higher than other thermoplastics and about equal to phenolic thermosets. Mold shrinkage was low and dimensional stability excellent. Water absorption was low and electrical properties were good. Lexan was basically transparent, though it then had an amber cast from impurities. On the negative side, Lexan parts were subject to stress-cracking and their resistance to solvents was poor.

GE set Lexan's introductory price at $2.50 per pound, about $1 more than nylon molding compounds, 3 times that of the acrylic plastics, and 8 times that of polystyrenes. This was a new engineering material, made by an expensive process, and GE believed its unusual properties justified an initial premium price.

Despite limited product and reproducibility difficulties, GE continued the broad sampling program. Lexan was very difficult to mold, and because special handling would be required for success in a molding shop, early Lexan samples were not shipped to prospects, but were hand-carried by two very busy GE market development representatives (Christopher and Richard J. Thompson). During nearly two years they stood by in customers' plants and coached the molding trials, so as to avoid application failures. The industry was familiar with some special techniques needed with the nylon thermoplastics of Du Pont and Celanese, but polycarbonate was much more difficult to mold properly and needed very high extruder cylinder temperatures (525°–600° F) and heated molds (170°–200° F). Different flow characteristics meant that Lexan needed bigger mold gates than was customary with nylon materials. Sealed, moisture-proof shipping containers were essential, and product that had adsorbed moisture from air exposure had to be heated and redried before molding.

In about nine months the sampling program had established 18 production jobs with a combined demand of 50,000 pounds per year. Four hundred major users were conducting serious evaluation programs. Demand for the product was exceeding the Pittsfield pilot-plant capacity, so GE purchased 100,000 pounds from Bayer in Germany and mixed it with equal parts of the pilot-plant product. This benchmarking revealed that Bayer polycarbonate at the time had superior clarity and moldability, serious challenges for GE's product development. Three professional market research studies had been conducted, of which Arthur D. Little's was particularly enthusiastic. Little concluded that Lexan properties were so superior to the commodity plastics that, if the cost could be brought down far enough, little else would be used. But Christopher declares it was more the market response from customers and prospects that gave GE confidence a large polycarbonate plastic market could be developed. Early applications showed the breadth of market interest: when molded of Lexan, a disc on a range switch did not melt, as it did when made from other plastics; card guides in IBM business machinery improved with Lexan's tight dimensional tolerances and stability; coil forms in TV components needed heat stability and good electrical properties; sight-tubes for GE coffee makers withstood very hot water and gained an early FDA approval; and light covers for the wings of supersonic aircraft could survive the thermal buildup of high-speed flight.

During GE's 1957 and 1958 market-sampling years, Bayer product was not being offered to U.S. customers, despite the presence of Mobay Chemical Company, a successful venture formed by Bayer and Monsanto in 1954 to offer urethane foam products in this country. By June 1959, the GE pilot plant had been expanded twice and was producing material for 150 different industrial applications. Lexan sales in 1959 were around 200,000 pounds.[4]

Lexan Project Transfer: A New Green-Field Plant

GE's CDO picked up the early technology from Dan Fox and the Research Lab, developed it to a usable product and process, and conducted the initial marketing. 1959 was the moment to bring Lexan into the Chemical Materials Department whose products were then phenolic resins and molding compounds, insulation varnishes, industrial paints, and purified magnesium oxide. The Lexan transition went smoothly, because both organizations were in Pittsfield, and Schubert became general manager of the Chemical Materials Department. Most of the Lexan development, pilot plant, and marketing personnel transferred from CDO to the operating department.

A next key decision would be how big, and where to build more plant capacity. No serious consideration was given to expanding in the Pittsfield area, as Ralph Cordiner's decentralization philosophy then discouraged new plants in existing GE employment areas. So an appropriation request proposed a new green-field plant, meaning built on undeveloped land, whose initial capacity would be 5 million lb/yr, to be located on the Ohio River at Mt. Vernon, Indiana. The all-solvent process would be used despite cost problems inherent in recycling pyridine and methylene chloride.

Corporate approval of this $6 million plant in early 1959 might be considered a leap of faith from actual 1958 customer sales of 25,000 pounds. But the demand curve from successful applications was rising rapidly, GE had a marketing lead over polycarbonate competition in the U.S., and the new thermoplastic showed extraordinary promise. And by this time Chemical and Metallurgical Division management had credibility and confidence from successful performances in silicones and in diamonds. Reed, who succeeded Robert L. Gibson as Division general manager in early 1960, later commented on the Lexan plant process and construction decision. "Even though we had a very poor expensive process involving pyridine which had to be recovered and recycled... If you really see yourself as first in the business... go ahead and take any process you have in order to get into production and make the product available to customers. Get your market position and in the meantime, go ahead with the development of the process and try to improve the economics of it. But get your market position first. This is the reverse of what we did in silicones. We had the process first, but no market and we learned from that experience."[7]

GE's plant-building announcement in June 1959 was followed a month later by a similar news release from the Mobay Chemical Company. Mobay, with headquarters in Pittsburgh, would build a polycarbonate plant of comparable capacity at New Martinsville, W. Virginia, adjacent to their urethane products facility.[9] Both new U.S. polycarbonate plants opened in the fall of 1960.

The Mt. Vernon start-up was facilitated by three years experience in the Pittsfield pilot plant, whose key managers, McCullough and Robert L. Hatch, transferred to lead the new facility. McCullough, the plant manager, had pioneered employment of nonexempt salaried technicians to operate the Pittsfield pilot plant, and he set up the same personnel system at the new location, thereby avoiding the usual distinction between an hourly paid and a salaried work force, which was the practice in most GE plants. This policy proved successful at the Mt. Vernon plant, where it contributed to a high level of employee involvement in the success of the business, along with an absence of feeling that union representation was needed.

Some GE Corporate Concerns

Although Lexan sales growth in 1960 and 1961 was in line with expectations, the product line showed larger losses than contemplated in the plant appropriation request. A yearly net income loss exceeding $1 million after developing a new plant site for a single product line might not have surprised an experienced chemical industry top management, but it drew attention from Fred J. Borch, newly appointed executive vice-president and heir apparent to CEO Cordiner, and some GE Board members. The Board committee pressed department management for a new commitment to a break-even date. John T. Castles, who succeeded Schubert as general manager when the latter accepted a higher-level position in large transformers under Gibson, and his staff sweated a plan update, which he and marketing manager Walter J. Dugan presented in February 1962. Both Castles and Dugan were silicone business alumni. Their plan committed to break even in 1965, but they advised that Mt. Vernon plant capacity would have to be substantially expanded by that time to meet the forecast sales. They also announced that the business must invest now in state-of-the-art molding and extruding equipment, plus a few industry-experienced people, for improved application development and demonstration in Pittsfield.

As an alternate to approval of this revised business plan, they said they were confident the polycarbonate business could be sold in its existing state for at least what GE had then invested. Borch and the Board members accepted the commitment, and subsequently approved the new capital expenditure requests. The Lexan managers were under continuous corporate pressure to reach profitability, and the product line did better than break even in 1965, even managing to report its first profit in the very month that Castles had been pressed to designate.

Market Growth

Within the GE Lexan management a serious discussion of pricing strategy continued. GE had gained market leadership and was the price leader. The issue was how quickly, and to what eventual level, the market prices could be brought down to stimulate the total demand and optimize the returns. As John C. Fisher of GE's R&D Center pointed out at the time in a paper, market demand for different plastics was strongly affected by price. GE's marketing communications suggested to customers and prospects that Lexan prices would come down as demand grew, and this did happen, even though the product line was in the red for several years.

The Lexan marketing program continued to create new customers, and sales grew rapidly. Communication by advertising, sales promotion, and direct sales calls established Lexan polycarbonates as the outstanding plastic for impact strength, heat resistance, and dimensional stability. Trade-show visitors were challenged to break a piece of Lexan with a sledgehammer, something almost none could accomplish, but all enjoyed trying. Sales grew to $1.6 million in 1961 and to $3.0 million in 1962. But most applications were a long, hard sell, because metal replacement redesign was usually required. Housewares manufactures were among the first to apply Lexan in their products, as in the novel GE electric carving knife. Power hand-tool makers became customers when shown that polycarbonate housings were an economic design for electrical safety requirements. Long Island Lighting Company became a proponent for Lexan street-light globes, even though the early Lexan transparency left much to be desired. Using the new Pittsfield extrusion equipment, GE developed transparent riot shields for police which could withstand a .45 caliber pistol bullet. Lexan hard-hat safety helmets proved superior to their metal predecessors. In the early years, substantial polycarbonate was also sold to competitor Mobay. By 1963 it was clear that demand would soon exceed the original 5 million lb/yr capacity, and plans were announced to triple the Mt. Vernon plant size. This was accomplished by early 1965.

Plant Process Changes

In addition to expanding the plant's polycarbonate capacity ahead of growing demand, many process changes were made in the 1960s. The plant startup had used purchased phosgene, whose shipping-container size was limited by federal regulation. Phosgene manufacturing technology acquired from the supplier and others was soon installed at Mt. Vernon. The phosgene process made carbon monoxide from coke, carbon dioxide, and oxygen, followed by reaction of CO with chlorine. The polycarbonate polymerization eliminated pyridine in 1963 by using a slaked-lime slurry for the HCl acceptor. Another backward integration was brought on line in 1967 with a 20-million-pound facility to make bisphenol A (BPA) from phenol and acetone. Technology for this step was purchased from Hooker Chemical and the Mitsui Petrochemical Company, and after some startup difficulties satisfactory operation was achieved. The internally produced BPA had far superior purity of the desired para isomer than that which could be purchased on the open market. And this improved purity made a dramatic advance in Lexan transparency.

Polycarbonate resin clarity and color became important qualities in the competition between GE and Mobay, and as these improved, numerous glazing applications opened up. Airplane windows were to become an important application. The Chemical Materials Department managers were disappointed with the pace of glazing market development through independent manufacturers, and decided in 1965 to integrate forward to GE-made Lexan sheet, glazing, film, and tubing. This added the complication of a final extrusion and forming step for these products at the Mt. Vernon plant, but it helped GE promote them vigorously. The value of Lexan glazing was greatly enhanced by the development of additives and treatments that increased its resistance to UV light and surface scratching. GE was ready with its own brand of "unbreakable" glazing when growing vandalism in U.S. cities stimulated demand for the product.

In 1965 GE obtained a license from Bayer to use a process in which BPA was dissolved in aqueous sodium hydroxide and then reacted with phosgene, with methylene chloride present as solvent for the polymer. This became known as the "interfacial" process, as the reaction proceeded at the interface between aqueous and organic solvent phases. After polymerization the aqueous salt solution and solvent phases were separated, the polymer washed and then precipitated from the solvent phase, and the solvent recovered for recycling. GE tested the interfacial process in the Pittsfield pilot plant and installed it at Mt. Vernon in 1968. The Lexan process equations (Fig. 6-3) are noted at the end of this chapter.[10,11]

The many process improvements, combined with the growing experience of the Mt. Vernon plant organization, finally yielded major manufacturing cost improvements and better margins, despite dynamic price reductions of nearly $1 per pound. Although Lexan sales had grown rapidly from the beginning, the cost burden of developing a new site and the time to make process improvements pushed the break-even year to 1965, eleven years after Fox's discovery, seven after commercial introduction, and five after the plant opening. But this was a considerable improvement over the GE silicone experience, which had taken fourteen years from Rochow's process discovery and seven years from plant opening to reach breakeven. By 1968 GE's Lexan products were a financial success.

Summary: Lexan Polycarbonates (1953 – 1968)

General Electric's success in polycarbonate plastics began with a novel polymer discovery and recognition of its unique properties. Although the original melt process was not satisfactory, an alternate synthesis produced an acceptable molding material in the lab and then the pilot plant. Key strate-

	$ (millions)								
	1960	1961	1962	1963	1964	1965	1966	1967	1968
Sales	$0.9	1.6	3.0	5.0	9.0	12.0	16.1	16.3	19.9
Net Inc.	$ na	(1.2)	(1.3)	na	(.07)	0.45	0.49	0.66	2.50

Early Lexan sales and net income.[4]

gy differences from GE's silicone experience were priority on *product* rather than *process* optimization, and speed to the market with early publicity and sampling to customers. The GE Lexan team was powerfully stimulated by knowing that strong Bayer competition was also working the field, so it rushed GE polycarbonates from the pilot plant to be first to reach U.S. customers.

Pilot plant expansions and purchases from competition supplied the market growth until the new Mt. Vernon plant came on stream. Then process and cost improvements, as well as volume growth, were factors in reaching profitability. Backward integration established both phosgene and bisphenol A manufacture, the latter providing dramatic increase in BPA purity and Lexan clarity. The polymerization process was continuously improved. A high initial price was reduced substantially to stimulate demand. Forward integration to make polycarbonate sheets, tubes, and glazing added manufacturing and marketing complexity, but allowed GE to promote these products more effectively.

Lexan polycarbonate was the strongest thermoplastic the world had seen up to that time, and with it GE launched into engineering plastics with U.S. market leadership over Mobay. In 1967 Bayer became sole owner of the Mobay Chemical Company, buying out the Monsanto interest after a Justice Department challenge concerning Mobay's formation.[12]

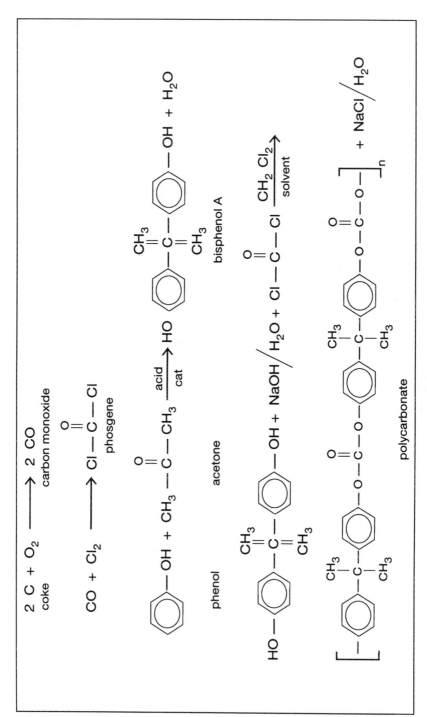

Fig. 6-3. Lexan polycarbonate process equations (1968).

CHAPTER 7

Noryl Thermoplastic:
1956–1968

Victory Snatched from Jaws of Defeat

As the Lexan polycarbonate development was getting underway in Pittsfield, Massachusetts in 1955, Allan S. Hay joined GE's Research and Development Center in Schenectady, New York. Hay was born in Canada, and gained his B.S. and M.Sc. degrees in chemistry from the University of Alberta; this gave him unusually extensive laboratory experience. He also had a Ph.D. from the University of Illinois. It was GE's growing interest in chemistry and the promise of some freedom to choose his field of research that attracted Hay.

Having a general interest in catalyzed oxidation of organic compounds, Hay explored new chemistry for converting xylenes to various phthalic acids. The latter were of interest to GE's alkyd resin and insulating varnish businesses. But the division's interest waned after the 1956 alkyd resin plant explosion and the subsequent decision to sell that product line. Hay had discovered a cobalt acetate bromide-catalyzed high-yield reaction to make terephthalic acid from p-xylene. Displaying some tunnel vision, Chemical and Metallurgical Division evaluators of that time predicted that terephthalic acid would not soon become commercially important. Hay and the GE patent attorneys did not file on his work, and were later chagrinned when a patent for bromine-assisted oxidation of p-xylene was issued to the Scientific Design Company. This process is used today for producing several billion pounds per year of terephthalic acid, and GE is a large purchaser.[1, 2]

Seeking a new route to bisphenols, Hay explored oxidation of phenol without interesting results, so he shifted to substituted phenols. In August 1956, Hay reacted 2,6-xylenol in pyridine with oxygen, using a copper catalyst, forming a viscous polymer. This was identified as a polyphenylene oxide, and it could be cast from solvent as a film. The reaction was this:

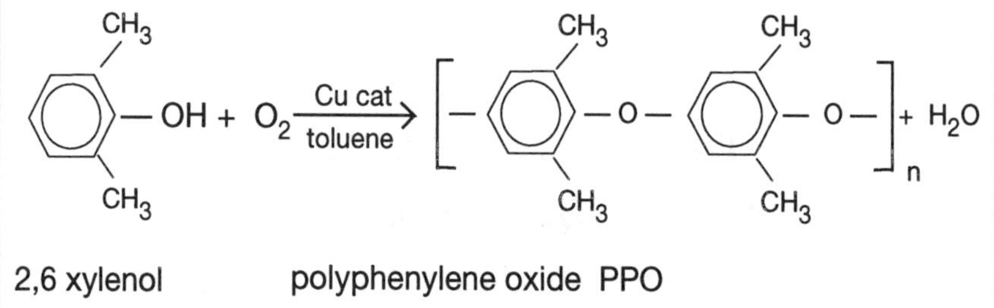

Fig. 7-1.

Here was new polymerization chemistry: it was not an addition, nor a condensation, nor a transesterification, but an oxidation. Over the next four years the R&D Center chemists expanded this research to see what it might yield in new polymers with unusual properties. Hay and his associates also explored variations of the reaction to simplify the process and to use less expensive solvents and catalysts. He filed for several patents, and in 1959 published on the subject in the *Journal of the American Chemical Society* and later spoke at an IUPAC meeting.[3,4]

As had happened in the Lexan development, the first polymer turned out to be the most promising. Polyphenylene oxide (PPO) thermoplastic was even more heat-resistant and hydrolytically stable than Lexan, and its electrical properties were excellent; but it was even harder to process in molding equipment because of its high melting point. Another serious obstacle to commercialization was inadequate monomer availability. At that time, the world's only 2,6-xylenol came as impure byproduct from coal-tar chemicals; so the supply was fundamentally limited at a modest level. Specially recrystallized material was available, but only in small quantities at $2.50 per pound.

Fox, Reed, and Gutoff Champion PPO

In those years (1959–1961), the Chemical Materials Department managers in Pittsfield, who were responsible for Lexan, objected to a new polymer program taking resources that might otherwise advance the Lexan effort, and to a product that might compete with polycarbonate applications. But Daniel W. Fox, who had returned from Lexan chemistry to manage advanced development in the Chemical Development Operation (CDO), believed that the unusual PPO polymer should be studied seriously, and he became its first champion. Funding was limited, so Fox personally scrounged some support from a GE fuel-cell program that might use a chemically modified PPO as an ion-exchange resin. He also promoted a contract from the Army Medical Corps for plastic surgical devices that could be autoclaved.[5] Fox touted the polymer to Reuben Gutoff, manager of CDO and an alumnus of the GE silicone business.

Division manager Charles Reed also became a PPO champion, stating: "Polycarbonates was draining a lot of money, but, again, with my experience in seeing how important it was to get into market development at an early stage, the first thing I did (after assuming position of Division General Manager) was to push very hard for them to take over the PPO work from the Research Lab. I said if Dow Chemical had had the development, with their interest in plastics, they would have already spent $5 million on it, which was big money in those days... my feeling was that you've got to move fast, at a very early date, when you have something that was as fundamentally good as I felt PPO was."[6]

In 1960, to fill a chemical engineering slot for developing PPO, Fox made a special, and successful effort to hire a visiting candidate from University of Illinois, John F. Welch, Jr. Welch's first assignment was building a pilot plant for PPO. Many years later, Welch would say of Fox at this interview:[7] "Here he had invented Lexan and was already pooh-poohing it and was off to make this new thing bigger and better. It takes a unique person to do that. Most people want to continue to do what they know best."

Monomer availability was first addressed by a world search that located a stockpile of impure 2,6-xylenol in the U.K.[8] Welch and his associates developed a liquid–liquid extraction purification and installed this process at Mt. Vernon, Indiana, to supply enough pure monomer for early commercial development. In parallel the R&D Center sought a new synthesis for high-purity 2,6-xylenol. By 1962, Stephen B. Hamilton had developed a satisfactory process by reacting phenol and methanol in a hot tube over magnesium oxide.[8] This process was scaled up in Pittsfield, and would later be installed at the PPO plant. Meanwhile, Hay and his associates developed

Fig. 7-2. Jack Welch, Allan Hay, and Reuben Gutoff at the Selkirk, N.Y., new plant site, which would make Hay's new polymer, PPO, precursor for Noryl thermoplastic. ca. 1965

another promising polymer by oxidative coupling of 2,6-diphenylphenol. This became known as P3O.

With these encouraging developments, Gutoff established a separate business venture within the CDO, the Polymer Products Operation, which built a 200,000 (soon expanded to 1 million) -lb/yr pilot plant in Pittsfield, and began market development of another new molding compound. GE announced PPO thermoplastic in October 1964. Yearly cost of the PPO effort had increased to around $1 million. Concurrently with expanding the pilot plant, and in spite of serious molding problems, Polymer Products Operation proposed building a 10 million-lb/yr green-field plant at Selkirk, NY. This $10 million project was approved in December 1964.[10] Sterilizable medical instruments made of PPO were cited as one probable application. The 500-acre Selkirk plant location south of Albany, NY, was chosen in preference to the less costly alternate of adding a new line at Mt. Vernon. CDO managers believed there was risk in moving PPO manufacturing far from the R&D Center and the Pittsfield pilot plant while the technology was still being developed; further, they were reluctant to become junior citizens at a Lexan plant site whose management regarded PPO as a competitive material.

GE corporate employee relations experts objected to the Selkirk location, which was near three unionized GE plants (Schenectady, Pittsfield, and Waterford), and they later attacked the policy of an all-salaried work force. But Reed and Gutoff prevailed, and the pioneering personnel practices that had proved successful at Mt. Vernon were adopted and worked equally well at Selkirk. Welch moved his personal residence from Pittsfield to the Selkirk area for a time during the plant startup.

PPO Fails in the Marketplace

As plant construction got underway, however, attempts to market PPO molding compound from the pilot plant were running into serious difficulties. This was a very different market reaction than the Lexan experience of 1957–1959. PPO was brittle and difficult to mold without degradation. The material was failing in application tests rather than being accepted.

The good news at this time was the patent situation. Hay's discoveries were clearly novel and would probably be covered worldwide. Hamilton's monomer process appeared to be patentable as a distinct improvement over prior patented art.[9] But worldwide patent coverage of a nonmarketable polymer and an unneeded monomer process would be a Pyrrhic victory. Hay's publications had attracted interest by chemists from a Dutch company, Algemene Kunstzijde Unie NV (AKU, now AKZO), who thought the PPO polymer might be developed as a fiber, a field in which they had experience. In subsequent discussions with AKU management, Gutoff negotiated formation of a joint-venture company called NV Polychemie AKU-GE (60% GE, 40% AKU), with headquarters in Arnhem, The Netherlands. This venture was the first GE chemical investment in Europe. Initial capitalization was kept under $100,000 so as avoid a time-consuming review for approval by GE's International Group Executive. Led by Leroy S. Moody, its first president, the company attempted development of PPO and P3O polymers as fibers, which was AKU's main interest, and PPO also as a molding compound, GE's interest. If fiber manufacture were to ensue, a joint company with an AKU majority interest would be formed.

The negative market reaction to PPO in the United States became a very serious crisis. The Selkirk plant was being built to make a thermoplastic that wasn't going to sell. Extraordinary effort was mobilized in Pittsfield and at the R&D Center to find ways of making PPO market-acceptable. It was then recalled that Edith Boldebuck at the Center had found that PPO and polystyrene were completely miscible.[11] Complete miscibility of differing plastic polymers is rare. Other supporters of a blend product were Robert Finholt, an alumnus of GE's major appliance plastics laboratory and of the Erie Works Lab, as well as Dan Fox, Popkin Shenian, R. P. Anderson, and Eric Cizek, to whom a key blend patent later issued.[12] A 25% PPO/75% polystyrene molding powder blend showed interesting properties, but it was still too brittle. Finholt placed a development contract with the major appliance lab, where many additives were tried, and a rubber polymer found whose addition in small quantity solved the brittleness problem.

Noryl Modified-PPO Blend: A New Engineering Plastic

The final blend, or alloy, trade-named Noryl, was a very different engineering thermoplastic from the one GE had set out to create. Instead of a polymer with some characteristics superior to polycarbonate, this Noryl alloy had physical properties intermediate between Lexan and acrylonitrile/butadiene/styrene (ABS) molding materials. While none of the properties was outstanding, the combination was unique, so the Polymer Products Operation management team, now headed by Welch, decided to market Noryl, not PPO. Gutoff and Welch broke the news to Reed that PPO would not be a successful product by itself, but that the Noryl blend looked good to some customers and might become a promising engineering plastic. The Selkirk plant would make PPO polymer as planned, and the polystyrene portion of Noryl, plus other additives, would be purchased, and all then mixed and compounded together.

The low-cost polystyrene content made it possible to price Noryl attractively on a cents-per-cubic-inch basis between polycarbonates and ABS. As Shenian put it, "Noryl offered a better combination of electrical and thermal properties than anything less expensive."[10] Noryl properties often qualified it to replace die-cast zinc parts. GE's Polymer Products Operation promoted Noryl plastic vigorously and the sales took off. Can opener housings, copy machine parts, and telephone handset parts were early applications. Flame-retardant grades became specified in aircraft interiors. The AKU-GE organization introduced Noryl in Europe in 1966. Golfer Welch would later say that Noryl was a "precisely placed 9-iron shot into an unoccupied region of the engineering plastics market."[10] GE's marketing effort fully established Noryl's value in that region. PPO polymer by itself never became a factor in the molding-material market.[13]

The Selkirk plant opened in December 1966. In 1967, the first full year of commercial operation, sales were nearly $2 million, and they continued to grow rapidly. In 1971, 6 years after Noryl's commercial introduction and 4 years after the plant opening, sales were $24 million and the product line earned its first profit. Sales had grown more rapidly than Lexan and the time to break-even was slightly shorter.

		1966	1967	1968	1969	1970	1971
Sales	$	1	2	4.5	8.7	16.6	23.8
Net income	$	(2.5)	(3.1)	(2.3)	(1.2)	(0.3)	0.9

Early Noryl Sales and Net Income[10]

$ Mills

Summary: PPO Spawns Noryl   (1956–1968)

Al Hay's polyphenylene oxide polymer discovery proved a commercial winner, not by itself, but as the key ingredient of Noryl blend. GE often referred to Noryl as "modified PPO." GE now had a second successful entry in the engineering plastics marketplace. Hay's patents and PPO technology, which GE did not offer to license, also became a major factor in leveraging the GE plastics position outside the United States.

The decision in late 1964 to move ahead with a $10 million plant construction for PPO polymer was exceptionally high risk, and the project faced failure until GE demonstrated the market utility of PPO blends with inexpensive high-impact polystyrene. Welch and his team showed great mental flexibility to make such a major change in product strategy. Noryl molding compounds were the first major blend of dissimilar polymers in the thermoplastics marketplace. GE would introduce several other important plastic alloys in the decades to come. (Note: I am using the terms "blends" and "alloys" synonymously, despite the fact that some thermoplastic mixtures are true solutions and others are not.)

Reed explained the logic of moving ahead in spite of the uncertainties: "You're constantly assessing what you would do in terms of capacity and location on a new plant. In the meantime you have some pilot-plant work going on the process itself and on the molding compounds... and you come to the point where you say, 'We know enough where we can take a gamble now on the first plant.' But it wasn't the first time that we did it. We did it with Lexan. We did it with silicones. So we had confidence. I had confidence."[6]

And the top levels of GE management had confidence in Reed, based on the steady growth and performance improvement of GE's chemical businesses.

CHAPTER 8

GE Engineering Plastics: 1968–1987

Headlong Growth to World Leadership

I n late 1967 a corporate reorganization rearranged the leadership respon-
sibilities for GE's chemical business. Charles Reed was promoted to
Group Executive, with continuing responsibility for chemicals, but also
for electronic components (transistors and tubes), appliance components
(small motors and controls), and medical systems. The previously assigned
conduit products line had been sold and the wire and cable and wiring
devices products were reassigned. The chemical products were now in two
divisions, with Reuben Gutoff heading the organization with responsibility
for plastics, silicones, laminates and insulating materials, while John T.
Castles led a division that included tungsten carbide tools, diamonds, and
magnets.

Welch Heads the Plastics Department

The Noryl project, now a going business under John F. Welch, was soon
combined with the Chemical Materials Department and that organization
renamed the Plastics Department. "Plastics" now meant a <u>material</u> rather
than a <u>molded part,</u> as in the original division product lineup. The name
change signaled the future focus of this small organization. In late 1968
Welch became general manager of this department, whose product lines
were now Lexan polycarbonate resins and sheet, Noryl molding compounds,

phenolic resins and molding compounds, and magnesium oxide insulating powder. The manufacturing locations were Mt. Vernon, Indiana, Selkirk, New York, and Pittsfield, Massachusetts, the latter also continuing as department and development headquarters. Plastics Department sales were around $40 million, less than GE silicones at that time, with the Lexan product line now profitable and Noryl startup losses about equal to Lexan profit. Both Lexan (around $20 million sales) and Noryl, ($4.5 million sales) were growing rapidly. Phenolic plastics were profitable, but not growing significantly.

U. S. Engineering Plastics in 1968 [1]

Thermoplastic molding compounds such as GE's Lexan and Noryl have been called "engineering" or "high-performance" plastics. The product class is not rigorously defined, but is distinguished from the higher-volume "commodity" thermoplastics such as cellulosics, vinyls, polystyrenes, acrylics, polyethylenes, and polypropylenes by higher melting points, higher strength, higher temperature resistance, and higher prices.

The first engineering plastics were nylon molding resins developed by Du Pont around 1941 and reintroduced after World War II in 1950. These polyamides (PA), which Du Pont trade-named Zytel, were a logical extension of Du Pont's pioneering polymer research.

In 1954 the Marbon Division of Borg Warner introduced an acrylonitrile/butadiene/styrene (ABS) thermoplastic resin family, which they trade-named Cycolac. While ABS resins are not always grouped with the engineering plastics, their tough properties and their lower prices than polyamides provided a big spurt at the time in new markets for premium thermoplastics. Cycolac plastics were rapidly adopted in a wide range of applications, including Western Electric's telephone housings. Monsanto, Dow Chemical, and U.S. Rubber also entered ABS manufacture.

As described in Chapter 6, GE and Bayer introduced polycarbonate resins (PC) in 1957–1958, the former in the United States and the latter in Europe. Polyacetal plastics (POM) were introduced in 1959 by Celanese (trade name Celron) and by Du Pont (trade name Delrin). GE expanded its line in 1965 with blended polyphenylene oxide (modified PPO), trade-named Noryl.

The engineering plastics industry had been launched. And with U.S. leadership in PC and modified-polyphenylene-oxide (PPO) types, GE had found its niche.

U.S. engineering plastics production in 1968 was less than 1% by volume of all thermoplastics:[2]

Thermoplastic	Approximate 1968 U.S. Production (million lb)
Polyethylene	4,570
Vinyls	3,220
Polystyrene	2,390
Polypropylene	930
ABS	510
Polyamides*	90*
Polyacetals,* polycarbonates,* PPO*	20*
All other	1,060
Total	12,790

*Engineering plastics

Welch Accelerates Market Development

Even though GE had U.S. market leadership in both Lexan and Noryl product lines, Welch, now in charge of the Plastics Department, saw the opportunity to accelerate growth by significantly increasing market development resources relative to technical effort. He cites this marketing increase as his first major initiative in the new job.[3] Without reducing technical effort, he stepped up advertising campaigns and found new ways to communicate the extraordinary strength of Lexan plastic, the values of Noryl, and GE's competence to assist the application process. Some great pictures and advertising copy came from the demonstration that showed that neither Bob Gibson nor Denny McClain, famous pitchers of the 1960s, could break a Lexan panel with their fastballs.

To help customer engineers design with plastic parts, Welch put technical assistance closer to the customers' place of business. In 1970 GE pioneered with a Detroit technical center to assist plastics-application engineers of the automotive and other Midwest industries. This local presence of GE Plastics was advertised on billboards along Detroit freeways as well as on the popular Bob and Ray daytime radio show, which reached a national audience. The tech center staff included application engineers as well as sales and district management. It was equipped with molding equipment and was backed up in Pittsfield by application specialists and more extensive equipment. Daniel W. Fox described the process, "You had to provide somebody to go hold the customer's hand, or have the customer in and hold his hand, and show him how to mold the stuff, how to modify his equipment and molds…you have to provide a technical service or your customer can't use your product."[4]

The Detroit tech center meeting place gave GE a strong identification with engineering plastics. In the process, the GE organization became familiar with auto company engineering practices, the influence of Society of Automotive Engineers (SAE) standards, necessary approvals, and customer personnel in many functions. Success of the initiative was easy to track by new application approvals from end users, followed by sales to their chosen molders. These facilities, today called Commercial Development Centers, are a worldwide hallmark of GE Plastics marketing programs, and they are replicated in Detroit, Atlanta, Pleasanton, California, and 10 other locations around the world, as well as at the Pittsfield and Bergen op Zoom, (BOZ), The Netherlands, headquarters locations.

World-Market Participation

Welch and the Plastics Department management team gave equal attention to market opportunities outside the United States. Because neither PPO nor P3O polymers had proved successful as fibers in the AKU-GE laboratory tests, Welch was able to negotiate a buyout in 1969 of the AKU share in the joint venture, renaming it General Electric Plastics N.V. (now GE Plastics B.V.). An important cadre of former AKU employees stayed with the new organization, including Jan Businck. BOZ became the General Electric Plastics N.V. headquarters in 1970, with a large land purchase for offices and future manufacturing. As would become the pattern around the world, plastics *compounding* was the first GE manufacturing activity at BOZ. Engineering plastics are usually compounded with fillers, additives, and colors to improve one characteristic or another before shipment to the customer. In the case of Noryl, about three-quarters of the resin content came from a supplier of high-impact polystyrene. The two resin powders were dry-mixed along with other additives and then forced through an extruder where the work energy helped melt the compound. The extruded molten plastic alloy strands were then continuously cooled to a solid under water, then chopped into pellets, and packaged or stored for bulk handling. Compounding of Noryl at BOZ for the European market began in 1970, and of Lexan in 1971.

Welch also negotiated a revised Bayer license that allowed sales of GE polycarbonates to begin in Europe a couple of years prior to expiration of the basic patents. GE Plastics was barely recognized in Europe at the time and was competing with Bayer, a well-known chemical giant. Welch recalls sending Kenneth Barr and Robert Kunze, two top marketing people, to Europe in those early days. GE's sales and applications people were very successful introducing Noryl and Lexan into Europe. Noryl was then a unique product, but Bayer polycarbonates (trade name Makrolon) had a previously unchallenged lead throughout Europe.

West Germany then represented over half the European market, so in 1973 Welch formed a German corporation, General Electric Plastics GmbH, at Ruesselsheim, to offer application help and sales representation. With manufacturing on the continent, plus local application expertise, GE showed it was totally committed to marketing plastics in Europe. Later the same year, technical and sales centers were opened in Manchester, England, and Evry, France. Centers in Italy and Spain followed in later years. Each country subsidiary was largely staffed with local nationals, often hired from the target end-user industries, such as electronics or automotive. GE polycarbonates market share in Europe approached 40% in 2 to 3 years, propelled by the new centers and aided by growth of the market. The GE organization succeeded by giving speedier responses to end-user and molder problems, and to questions of how to design with the new plastics.

In Japan, during 1971, Welch arranged a sales and compounding joint venture called Engineering Plastics Ltd., 50% owned with Nagase, a Japanese distribution company with experience selling Eastman Kodak products. This organization, led by Hideo Sato, proved effective in compounding and marketing all the GE engineering plastics for Japan and the Far East for many years. A compounding and marketing joint venture was also created with M. E. Hogg Australia PTY, Ltd., in 1972.

With minor exceptions, Gutoff and Welch chose not to use the services of International General Electric during this world expansion for engineering plastics, either for pooled sales representation or affiliate company relationships. General Electric in these years was in a slow transition outside the United States, from components managed by International General Electric (IGE) to U.S.-led separate product businesses with world responsibility.

Valox Polyester Plastics

GE's growing competence and success with engineering plastics provided insight for the next major addition to the product line. The GE Plastics managers were no longer interested in branching out into fiber polymers or new thermoplastic recording tapes, each of which had been studied by the Chemical Development Operation (CDO). A strategic focus on high-performance molding materials had developed with the success of Lexan and Noryl, each of which originated from a new technology. The next product would come from a perceived market need. According to Dan Fox at CDO,

"Our marketing people kept telling us that now we had two *amorphous* polymers. They had great thermal stability and everything else, but they dissolve in everything."[4] So CDO personnel set out to find an appropriate *polycrystalline* molding resin with excellent solvent resistance as well as good high-temperature performance. For automotive applications, such a new plastic would allow GE products to get under the hood of cars and other gasoline-affected locations, filling a major gap in the existing product spectrum.

Two polymers were seriously studied, one from a Rhone-Poulenc license, a promising high-temperature and chemically resistant material. This failed on scale-up for processibility reasons. The second was a plastic that had recently been introduced by Celanese. Analysis indicated it was a glass-filled polybutylene terephthalate (PBT) polyester. Despite the 1969 commercial introduction of this "Celanex" product, the polymer technology was old and basic patents had expired. Welch asked Fox if GE could make the stuff and Fox pointed out that there were no patent barriers. "Go make some," said Welch.

Fox and Pim Boman made several pounds in the laboratory and the properties looked excellent. Welch then told them "Buy a kettle. I'll get the signature."[4] So GE developed a PBT molding compound, scaled up a suitable process and introduced it to the market. GE trade-named its PBT polyesters "Valox" and built a 2.5 million pound per year pilot plant in 1971.

Reed, who was then responsible at the group executive level, described the reasoning for going ahead without a proprietary technical position: "I've always said that we had to have some "unfair advantage," and the only unfair advantage we had in Valox was our previous bank of experience in developing the new polymers commercially. I felt that we had such an outstanding marketing and development operation at that point, under Welch, that we could take on Valox even though we had no advantage in monomers or in process chemistry at that point. But what we had was a powerful marketing organization."[5]

Speedy action was characteristic of Welch-led organizations. Valox was introduced to the United States market in 1972 and generated $1.2 million sales that year. Customer contacts were simple because GE sales personnel were already calling on the molders with Lexan and Noryl, while GE market development specialists were calling on end users. Pilot-plant capacity was stretched to 10 million pounds and sales reached $5.7 million in 1973 and $15 million by 1974. A major production scale-up for PBT was built at the Mt. Vernon plant in 1974, this time without the expense and delay of developing a new location, as had been the case for Lexan and Noryl. Despite many PBT competitors, GE gained a leading market share, followed by Celanese. The Valox project broke even in 1976, with sales of around $30 million, four years after introduction. Contributing to the rapid sales growth was GE's global marketing reach, with Valox introductions in Canada in 1972, Australia in 1973, and Europe and Japan in 1974.

The more rapid sales growth for Valox than either Lexan or Noryl (Table 8-1) also yielded a shorter time to breakeven. This reflected not only the product's suitability for applications but also GE's growing effectiveness for converting market potential into sales. It was a convincing demonstration of the company's competitive competence in engineering plastics. Valox's excellent solvent resistance and good high temperature performance expanded GE's served-market opportunity in engineering plastics nearly 40%.

Table 8-1 Comparative Sales Growth Patterns ($ millions)[6]			
	Year 1	Year 5	Year 10
Lexan	$ 0.07	3.0	13.6
Noryl	$ 0.12	4.5	63.5
Valox	$ 1.2	30.0	81.5

Plastics Product Line Proliferation[6]

GE now had three thermoplastic polymers: an aromatic polycarbonate (Lexan PC), the polyphenylene oxide blended with polystyrene (Noryl modified-PPO), and a polybutylene terephthalate (Valox PBT). Lexan had high-impact strength and temperature resistance; Noryl was easily molded, cost less than PC, and surpassed ABS in some properties; and Valox had outstanding oil and solvent resistance. The performance range of these three engineering plastics was broader than that from any competitor in the world.

During the early market years for each, a single polymer grade was optimized for finished-properties balance and for moldability. While the production managers of GE chemical plants might hope that one polymer would be the sole demand on their complex process (the original Lexan plant program assumed one polymer grade and five colors), this was never to be. In response to different market needs and competitive offerings new polymer grades were developed in each line, as well as a plethora of grades compounded from various fillers, additives, and colors. Flammability resistance became an important characteristic for every engineering plastic, and improvements were sought both by polymer modification and by additives. Foamed plastics were useful in certain applications, so each of the polymers was offered in foamable grades.

In Lexan, new versions included flame-resistant Lexan, glass-fiber-reinforced Lexan, mold-release Lexan, supertough Lexan, and a version for blow molding. Chemical-resistant, E-Z flow, and foamable grades followed. The 1980s saw even greater proliferation, including new structured products in film, very high and very low molecular-weight polycarbonates, and a whole new Lexan copolymer family known as polyphenylcarbonates.

Lexan sheet and glazing was continually improved by better clarity, greater UV resistance over long periods, and improved scratch resistance. Lexan glazing became important in the United States, Europe, and Japan. It was a major product of the Mt. Vernon and BOZ plants, and GE also set up glazing joint ventures in Italy and in Japan to serve these markets effectively.

Noryl plastics had become a "family" by 1970. One grade had "good dimensional stability to 265°F." Glass-filled Noryl had "excellent mechanical strength and heat resistance to 300°F." Noryl eventually included chemical, fire, and UV-light-resistant grades, plus gloss and antistatic product variants.

In like manner the Valox line by the mid-1980s comprised seven families, including foam, high-temperature, and electromagnetic interference shielded grades.

Organization and Market Development

During the 1970s expansion of GE's three engineering plastics product lines, the U.S. product development chemists, who were originally located in Pittsfield, gradually moved to labs at Mt. Vernon and Selkirk. Longer-range product development continued in Pittsfield under Fox, while Allan S. Hay led the substantial chemistry research effort at the Corporate R&D Center (C R&D).

In sales and marketing, even though GE now had Lexan, Noryl, and Valox in a single division, Welch and later managers maintained competition between the products by having separate U.S. sales and market-development forces for several years. Pooled selling with a single national sales force was set up in 1974, but product specialization continued among the market-development representatives, who called on end users. This reflected the different marketing approach needed to gain customer approval for a new plastics application compared to that required to compete for orders once GE products were approved, and sophisticated measurements of this market-development effectiveness were developed. Targeted new applications were tracked by industry, end-use customer, molder, pounds per year potential volume, and finally, orders received. As John D. Opie, who once managed the U.S. Lexan business and is now a GE Vice-Chairman puts it, "GE is great at measuring things, and we really measured the results of market development effort."[7]

The success pattern found with Lexan and later with other plastics was to seek out applications in a range of markets and then, over time, expand those niches and add new ones. GE actually analyzed the housewares industry by subcategories of kitchen electrics, personal care, power tools, floor care, lawn and garden, major appliances, food service, and liquid handling. The proliferation and expansion of application niches continued through the 1970s and 1980s, and GE marketing people came to think of markets and life cycles. In order to keep total sales growing, the creation of new niches at the start of their life cycles needed to go forward faster than the sales losses in segments toward the end of their cycle.

The automotive sector was a special market that developed more slowly than appliances, consumer electronics, computer, and other business-machine applications. But each successful automotive application generated large demand for the chosen plastic. The first automotive applications for Lexan were for lighting parts, followed by interior items, and finally for bumpers and similar body parts. As the auto industry focused on increased fuel efficiency in the 1970s and 1980s, weight reduction became a decisive plastics value over metals, in addition to cost improvement. A world energy crisis in the 1970s drove up raw material costs and selling prices of engineering plastics; on the other hand, however, the energy content and price increases for metals were even greater, so engineering plastics advantage over metals increased further.

The market growth for engineering plastics was closely tied to continuous improvement of the product technology. GE estimated in the 1970s that one-third of total plastics sales 5 years out would come from product grades that were yet to be introduced. This new-product growth contribution was similar to that shown earlier in silicones (Chapter 3). The GE Plastics marketing organization became skilled at matching their product capability to the customers' application needs, using the most appropriate molding or extrusion equipment available. Of course, the proliferation of grades posed a monumental challenge in plant scheduling and quality control.

Growth of Sales and Capacity

Engineering plastics was a rapid growth market in the 1970s, GE was participating in a large part of it, and was growing faster than the industry by aggressive worldwide market development and product-line enhancements. By the mid-1970s GE's Lexan, Noryl, and Valox combined sales had become larger than either the company's silicones or tungsten carbide products. From around $100 million sales in 1972, GE engineering plastics sales grew to over $700 million in 1980. In that year Lexan accounted for roughly $325 million, Noryl $300 million, and Valox about $55 million.[6] The growth required continuous plant expansion investments, which increased GE capacity in engineering plastics to 600 million pounds per year by 1982, and to over 1 billion pounds per year by the second half of the 1980s. Major 1970s polycarbonate resin capacity additions included Lexan polymer and sheet at Mt. Vernon in 1973, and 80 million lb/yr Lexan polymer and sheet products at BOZ in 1974. In 1977, Valox PBT capacity was increased to 130 million lb/yr, with a new continuous polymerization facility at Mt. Vernon. In 1978, PPO polymer manufacturing capacity was brought on stream at BOZ. In the 1970s and 1980s GE Plastics grew much faster than the company overall and required continuing major plant investments.

Process Improvements

The continuing growth of Lexan polycarbonates stimulated process improvements along with capacity expansion at the Mt. Vernon and BOZ plants. During the 1970s GE successfully converted the interfacial polymerization for polycarbonate from batch to continuous operation, lowering costs and increasing capacity further. This was the *fourth* polymerization process since the plant opened. Additional capacity for superpure bisphenol-A was installed at Mt. Vernon. The purchased raw materials now included both sodium hydroxide and chlorine, while the polycarbonate process was yielding a byproduct stream of aqueous sodium chloride, so GE added a backward integration cost improvement in 1976 by installing electrolytic cells to convert the byproduct brine solution back to caustic and chlorine.

Phenol Integration

GE's rapidly growing polycarbonate and polyphenylene oxide plastics each used phenol raw material, so GE became a very large user. With consumption at around 200 million pounds per year in the mid-1970s, GE was the largest merchant user of phenol in the United States. Backward integration to make phenol became a logical study.

British Petroleum and Hercules in the 1950s had developed a two-step oxidation of the petrochemical, cumene (2-propylbenzene), to phenol and acetone (Reaction 8-1). Cumene was readily available from World War II use in high-octane airplane engine fuel. This phenol process was available via licensing, plant design, and construction from the M. W. Kellogg Company.

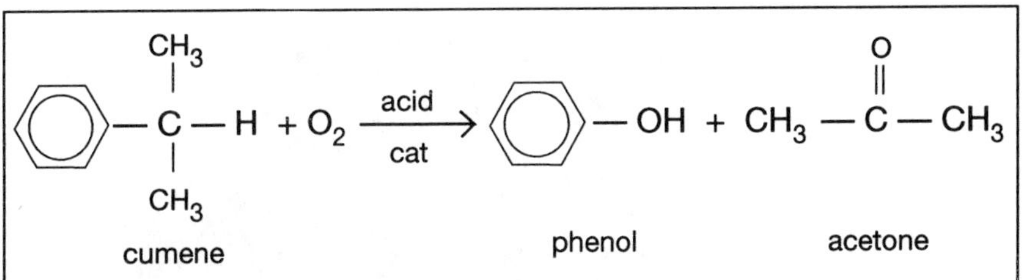

Fig. 8-1. Phenol process from cumene

For GE, which had a concurrent demand for part of the acetone to make bisphenol-A, the cumene process was very suitable. The management issues were timing and a satisfactory return on a very large plant investment. As GE seriously studied making phenol, suppliers gave ground on price from time to time. Finally GE approved a $90 million phenol plant of 400 million pound capacity in 1977; this plant went on stream at Mt. Vernon in 1980. After this process addition, GE's raw materials for polycarbonate plastics were coke, oxygen, carbon dioxide, and cumene, along with some recycle makeup of methylene chloride, sodium hydroxide, and chlorine; and GE became a merchant acetone seller. And as had happened earlier with bisphenol-A integration, GE found it could make purer phenol than the purchased grade and thereby improve the downstream plastics processes.[8]

Another backward integration opportunity presented itself when R&D Center scientists developed a unique process for making 1,4-butanediol, one of the monomers for the Valox PBT polymer. Up to now GE has not chosen to manufacture this diol, but has relied on supplier awareness of its process capability to negotiate fair pricing of the intermediate.

Promotions in the Organization

By 1972 GE had promoted Reed to Senior Vice-President on the corporate staff, and he was succeeded by Gutoff. The latter was soon promoted again to head strategic planning for the corporation. Welch advanced, first to Division Vice-President level in 1972, with responsibility for all the chemical operations, and in 1974 to Group Executive level. His group scope included all the plastics, the other chemical and metallurgical products, and three other divisions. Instead of moving to GE corporate headquarters in Fairfield, Connecticut, Welch got permission to keep his office location in Pittsfield. The plastics business had been elevated to division status in 1973, with Donald Debacher, who had managed the Lexan pilot plant, was later manager of the Mt. Vernon plant, and then headed GE Silicones, as its head.

Fig. 8-2. John F. Welch, Jr., Chairman-elect, and Reginald H. Jones, Chairman and Chief Executive Officer, 1981.

The rapidly growing Plastics organization gained a company reputation for freewheeling and independent thinking, and some GE observers describe the special culture as a "wild-west atmosphere." GE Plastics leadership encouraged internal competition and conflict, though intense concentration on what it took to gain and hold customers was a shared common cause. The organization saw itself as a high-flying success, and consequently some corporate administrative pronouncements or suggestions were ignored or taken lightly.

Reginald H. Jones, who succeeded Fred J. Borch as GE's Chief Executive at year-end 1972, recalls the plastics sales growth and the continuing requests for plant expansions. He supported the large investment program because he was impressed by the soundness of organization's strategic plans and the performance record on prior plant approvals. Company interest in plastics growth had increased after GE found it necessary, because of heavy losses, to sell its computer equipment business (1970), and as prospects for nuclear power equipment slowly faded. Jones later insisted that a reluctant Jack Welch leave Pittsfield at year-end 1977 to become Sector Executive (a newly created step up from Group) in charge of all GE's consumer products plus the GE Credit Corporation.[9]

Welch was promoted to Corporation Vice-Chairman in 1979 (in which position he again had executive authority over GE Plastics). He was designated GE's next Chief Executive in December 1980, and he succeeded Jones in that position in April 1981.

Ultem Polyetherimide Plastic

GE's engineering plastic polymer families—Lexan, Noryl, and Valox— provided broad coverage for high-performance applications requirements. Nonetheless R&D Center chemists had an unwritten goal to synthesize polymers whose high-temperature performance might exceed PC or PPO, and which would flow freely enough, without degradation at melt temperatures, to be molded like other engineering plastics.

Just as the Pittsfield Chemical Development Operation was beginning to scale up Valox (1969), Joseph G. Wirth at the R&D Center explored a polymer synthesis by displacement of aromatic nitro groups with various sodium phenoxides. Wirth transferred to GE Silicones as R&D manager, but Al Hay believed the potential polymers were sufficiently interesting to expand the program to a 5-year effort in which Howard Relles, Tohru Takekoshi, James Webb, Frank Williams, Newell Cook, and Hay himself made important contributions to both new monomer and polymer synthesis. Of many possible polyetherimides the eventual polymer design was prepared from bisphenol-A, 4-nitrophthalimide, and m-phenylenediamine. The polymer structure (Formula 8-3) is on the following page.

GE Plastics management was not enthusiastic about starting a commercial polyetherimides program, and the major skeptics included both Fox and Welch. Hay, manager of the project at C R&D, describes the internal obstacles:[10] "We had a material with interesting properties and then we showed we could make the material; then the critics said, "well, it's so complicated you're never going to be able to manufacture in the plant." So you demonstrate the process can be utilized...then of course you have people saying, "well, you've got the stuff, but you never really can mold it." Then you got all that out of the way and it comes down to, "Well, you'll never be able to sell the stuff."

Plastics Division Vice-President Donald Debacher, after a team visit from Arthur M. Bueche (Vice-President, Research and Development), Hay, A. R. Gilbert, and Relles, finally agreed to begin a joint evaluation program. The commercial evaluation began in 1974 under Jack Lidstone, a veteran of both plastics and silicones marketing, while the R&D Center continued to scale up the complex syntheses. The product under development was designated X-76, as a 1976 commercial introduction was anticipated. This date was not

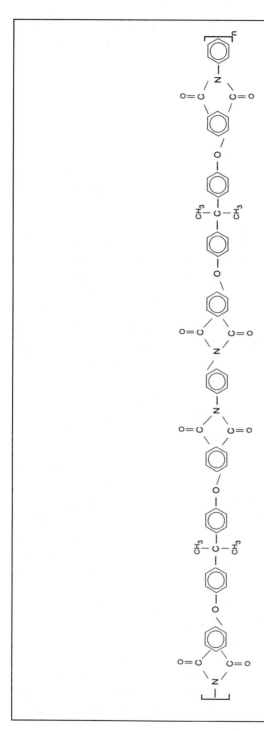

Fig. 8-3. Polyetherimide PEI

met, but the properties of Ultem 1000 eventually proved to be as exceptional as those anticipated from the polymer structure.[6] The Ultem polymer has greater strength than polycarbonate, and its heat distortion temperature is 140°F higher than Lexan. It has excellent thermal oxidative stability, and continuous use at 330°F (or even higher) is feasible. Electrical properties are exceptional over a wide temperature range. Flammability resistance of the polymer, even without additives, in various tests is outstanding. Ultem molds easily without degradation.

Wirth estimates that the 8-10-year R&D effort to bring this technology to the manufacturing point cost $25-50 million.[1]

A $1 million, 50,000-pounds-per-year pilot plant came on-stream in 1979, and feedback from the market was sufficiently favorable that a $20 million, 3-million-pound plant was installed at Mt. Vernon in 1982, using a five-step process. The process complexities required a price of around $6 per pound, and the price today still exceeds $5 per pound. But applications for this exceptional high-temperature plastic were found in several market niches, including surgical staples, microwave cookware, circuit boards, and aircraft components. GE authorized a major plant expansion costing $75 million to begin operations in 1986.[6]

Two other product ventures were explored in the 1970s. One, named XB-1, originated at the R&D Center, and was similar to Lexan but with superior flammability resistance. This was dropped when it proved costlier than Lexan and when Lexan's flammability resistance was significantly improved. The other, an epoxide initially called X-118 and later Arnox, would complement Valox. Arnox was produced in pilot-plant quantities by 1978 and marketing programs were begun with a target of 8 million pounds per year. GE gave Arnox substantial publicity, but the expansion did not take place because technical problems forced termination of the project.

A New Plastics Leader

Just before leaving Pittsfield in 1977, Welch recruited Glen H. Hiner to be head of GE Plastics European operations. Hiner's early GE career had been in electrical businesses and he was heading a motor department in Ft. Wayne, Indiana, at the time. Hiner relates that in November he flew from Chicago to meet his new associates for the first time at a sales meeting in Switzerland. A BOZ manager greeted him with the news, "There's been a phosgene release at the plant."[11] Hiner was gratified to learn that there were no casualties and that emergency measures had quickly controlled and detoxified the area. He directed that the BOZ polycarbonate process be shut down pending thorough investigation and planning for new safeguards. A completely new enclosure design and safety procedures had to satisfy not only GE management but also the very concerned Dutch environmental control authorities. The BOZ polycarbonate process remained shut down for months while Lexan shipments from the Mt. Vernon plant supplied the European demand. The extraordinary phosgene enclosure system was also added at the Mt. Vernon plant. In late 1978 Hiner returned to the United States to become Vice-President in charge of the Plastics Division, reporting to Group Executive Charles R. Carson. Hiner's leadership of GE Plastics continued for the next 13 years.

Xenoy Blend Automotive Bumpers (PC and PBT)[1]

After noting that the Ultem development had taken 10 years and that it was the first important engineering plastic since Valox, Hiner proclaimed, "the time and cost to develop a new polymer is prohibitive. Our growth will come from alloys, copolymers, and new applications."[6] GE's product emphasis in the 1980s became either variants and blends of existing products or complementary materials acquired from other companies.

Around 1978, modification of Lexan polycarbonates with elastomers improved the stress-cracking characteristic, and it became possible to consider car bumpers made of this material. But gasoline resistance had to be improved. This challenge led GE Europe scientists in 1980 to elastomer-modified polycarbonate/polybutylene terephthalate (PC/PBT) blends, designated Xenoy CL 100. This blend of immiscible polymers had medium heat resistance, the high ductility and dimensional stability of amorphous polymers, and the stress-cracking and gasoline resistance of semicrystalline polymers. The application development for self-colored, unpainted bumpers of the European Ford Sierra was pioneered by engineers of GE Plastics B.V. from BOZ several of whom worked in the Belgian Ford plant for months to assure application success.

Additional requirements for painted U.S. car bumpers, which had to pass a 5-mph crash test, were met in 1984 with Xenoy CX1101 blend, and this material was used to make bumpers of the Ford Taurus and Mercury Sable. By the late 1980s Xenoy blend was being used on roughly 35 car models around the world.

Noryl GTX Blend (PPO/PA)[1]

Automotive industry interest led to investigation of other nonmiscible polymer blends. University research had demonstrated how judiciously selected block and graft copolymers can stabilize the dispersion of truly incompatible polymers, those that do not form a true solution after melting

106

together. The resulting blends can have enhanced properties. In 1984 GE's Europe area chemists applied this technology to blends of polyphenylene oxides and polyamides (PPO/PA), creating a new thermoplastic family, GE trade name Noryl GTX. The blend takes advantage of polyamide chemical resistance, processability, and paintability, while adding heat resistance, dimensional stability, and low-temperature toughness through PPO.

This balance of properties has led to mass production of some on-line painted automotive special panels on some Peugeot and BMW car models.

Engineering Plastics: ca. 1981[12]

A 1982 *Chemical and Engineering News* survey article documented the leading engineering plastics as polyamides and polycarbonates at about equal volume, followed by modified polyphenylene oxides, polyacetals, polyesters, and polysulfones.

Engineering Plastics	1981 U.S. Consumption (mill. lb/yr.)
Polyamide	275
Polycarbonate	250
Modified polyphenylene oxide	125
Polyacetal	95
Thermoplastic polyesters	55
Polysulfone	10
Polyphenylene sulfide	7
Others	1
Total	818

Data in the article also showed that in the United States GE had in place more capacity than the next two competitors combined, though in different plastics. GE did not make polyamides or polyacetals, while Du Pont and Celanese were not in polycarbonates or polyphenylene oxides.

Major U.S. Producers: Engineering Plastics

(1981 U.S. Capacity, mill. lb/yr)

General Electric: 605

Polycarbonate (265); polyphenylene oxide (200); thermoplastic polyester (130); polyetherimide (10)

Du Pont: 288

Nylon (160); acetal (80); thermoplastic polyester: (25); fluoropolymers (20); polyimide (3)

Celanese: 235

Acetal (125); thermoplastic polyester (70); nylon (40)

Mobay (Bayer):   110

Polycarbonate (90); thermoplastic polyester (20)

Monsanto: 70

Nylon (70)

The world engineering plastics production, not including ABS, totaled around 1.75 billion pounds in 1981, a tonnage niche less than 2% of all thermoplastics, though a much higher percentage of the dollar volume. World production capacity for all thermoplastics totaled around 140 billion pounds:

World Thermoplastics	Capacity (mill. lb/yr)
Polyvinyl chloride	32,430
Low density polyethylene	31,702
High density polyethylene	18,071
Polypropylene	15,631
ABS	4,374
LLD polyethylene	2,669
All other (incl.1,750mil. engrg. plastics)	18,755
Total	140,123

On a volume basis, the world's leading producers of thermoplastics in 1981 were estimated to be the following:

Company	(mill. lb/yr)
Dow Chemical	7,900
BASF	7,600
Shell	7,420
ICI	6.680
Hoechst	6,400
Union Carbide	5,310
Montedison	4,430
Solvay	4,360
Du Pont	3,760
British Petroleum	3,510

It is evident from these tables that General Electric's engineering plastics output was trivial tonnage compared to the producers of commodity thermoplastics. But with a high market share in premium-performance plastics, GE had gained a very significant dollar volume and earnings position by sticking to its "niche strategy."

Exit from MgO and Phenolics

GE stopped manufacturing purified magnesium oxide in 1981 and thermosetting phenolic plastics in 1982. GE had produced MgO insulation powder, which made possible the metal-sheathed heating elements of many appliances, in or near the Pittsfield Works since the 1920s. Phenolic resins and molding compounds had been produced in the Pittsfield Works since 1929, originally for internal use in laminates and molded parts and after 1947 for external sale as well.

Continuing World Expansion[13]

Growth of GE's plastics sales in Japan and Asia prompted Hiner to renegotiate the Engineering Plastics, Ltd. (EPL) joint venture with Nagase in 1980, increasing GE ownership to 51%, so as to have clear management responsibility for decisions on how to serve customers in Asian countries. The earlier investment in Australia was increased to 100% in 1981, and was renamed General Electric Plastics Australia, Pty. Ltd. In late 1983 the GE Plastics operations were elevated to Group status, with Hiner becoming a senior vice-president.

To supplant PPO polymer exports from the Selkirk plant to Japan for compounding by EPL, Hiner formed GEM Polymers Ltd. with Mitsui

Toatsu and Mitsui Petrochemicals. GEM Polymers built a 2,6-xylenol plant at Osaka to supply monomer for a new PPO polymer plant at Takaishi, each coming on stream in 1984. A new wholly owned subsidiary, GE Plastics Japan Ltd., headed by Hideo Sato was formed to provide management services to the Japanese affiliates and to develop new ventures. GE Plastics management was pleased that these Japanese plants could be built on a shorter time cycle than was possible in the United States, and that production quality and output met and sometimes exceeded that of similar plants in the United States and Europe. Building on these successful projects, another was created with Mitsui Petrochemicals in 1986 to make high-purity BPA as feedstock for polycarbonate plastics, and a Japan PBT and a polycarbonate polymer venture with the same partner followed in 1988.

A major compounding plant for Noryl was built at Campinas, Brazil, in 1985 to serve computer and business equipment customers.

A second U.S. polycarbonate plant site had been selected, but construction was deferred in 1982 when the recession reduced demand. But as automotive applications continued to grow and a new PC application for CDs and CD-ROMs developed, the largest single capacity expansion in GE Plastics history began in 1985 at Burkville, Alabama. The $325 million plant, which came on stream in 1987, was GE's third for polycarbonate polymers, following Mt. Vernon and BOZ.

Under pressure from Welch to improve the Group's return on investment, Hiner negotiated an unusual sale and leaseback of GE's Mt. Vernon phenol plant to corporate investors, including Citgo, which was interested in being the plant's cumene supplier. He made a similar deal with Huntsman Chemical with respect to high-impact polystyrene facilities installed at Selkirk. Such transactions typically improve cash flow and return on investment, but reduce net earnings.

Investment in technical centers had also continued around the world. In 1979 the Detroit center was relocated and expanded to include computer-aided parts and mold design capability. In 1980 a new facility in Japan was located at Gatemba, near Mt. Fuji. GE opened centers in Toronto, Canada, in 1983, at Norcross, Georgia, in 1984, and near Milan, Italy, in 1985. A new Europe Technology Center was built at BOZ, and the Pittsfield application center expanded substantially in 1984, receiving recognition as "Laboratory of the Year" by *Research and Development* magazine.

GE Plastics also entered several joint ventures in the intermediate application of plastics for special end uses. AZDEL, the largest of these, was created in 1986 with Pittsburgh Plate Glass. AZDEL produces laminate composites of various glass fibers with (mainly) polypropylene, which can then be formed to intricate shapes by preheating followed by stamping processes. AZDEL technology has not used as much GE-manufactured polymer as had been anticipated, however.

New Polymers from GE

A weather-resistant resin terpolymer developed by Stauffer Chemical became available for GE purchase when Stauffer was first acquired by Chesebrough-Ponds and then spun off to ICI. GE trade-named the acrylic/styrene/acrylonitrile (ASA) plastic Geloy, and built a manufacturing facility at the Selkirk, New York, location, in 1985. Exterior construction paneling is a large application target for this weather-resistant polymer.

A family of thermoplastic elastomers, or "soft plastics," was developed in 1985 and trade-named Lomod. These are block copolymers of polybutyl-terephthalates and polyethers.

GE also added a polyphenylene sulfide product group named Supec in 1987, purchased for resale from Japan's Toso Susteel company. The polypheny-lene sulfides are crystalline, high-temperature, and solvent-resistant polymers.

Competition in Engineering Plastics

Mobay (which would change its name to Bayer in 1995) expanded with the U.S. polycarbonate market by building a new plant at Baytown, Texas, which came on stream in 1980. Mobay also began manufacturing a PBT polymer trade-named Pocan and established a plastics compounding facility at Newark, Ohio. Bayer built a major new PC plant at Brussels, Belgium, in the mid-1980s and maintained Europe leadership in the polycarbonates, closely followed by GE. Also in Europe, Bayer introduced Bayblend, a mixture of polycarbonate and ABS that was competitive with Noryl in some applications.

After some years of pilot production, Dow Chemical became the third U.S. polycarbonate producer by building a 30-million-lb/yr plant at Freeport, Texas, in 1985. Du Pont developed and introduced a new nylon-based thermoplastic line named Bexloy for certain automotive panels. Borg Warner targeted Noryl applications with a PPO copolymer trade-named Prevex and installed a 110-million-lb/yr plant in Mississippi.

Teijin Chemicals and Mitsubishi Gas Chemicals had each entered polycarbonate production and became the leading producers in Japan. Teijin specialized in optical-quality resins specially tailored and packaged for CDs and CD-ROMs. Polycarbonate use in this large application depends on extreme polymer purity, dimensional stability through moisture and temperature cycles, plus a rapid molding cycle for economic fabrication. Both Asahi Chemical and Mitsubishi Engineering Plastics in Japan began making PPO polymer in competition with GEM Polymers.

Summary: GE Engineering Plastics: (1968-1987)

By 1968 GE's Chemical and Metallurgical Division had established two small beachheads among premium-priced engineering plastics: Lexan polycarbonates and Noryl modified polyphenylene oxides, with combined sales around $25 million. Although corporate management was looking for growth product lines, in 1968 there was no realization at the top of General Electric that engineering plastics had potential comparable to three big ventures the company was then counting on for growth: (1) nuclear power equipment, (2) commercial aircraft jet engines, and (3) computers. (Of these, only commercial aircraft jet engines would succeed.) Nor was there any thought in Pittsfield in 1968 that plastics would ever rival the importance of GE's large transformer business.

Lexan and Noryl each stemmed from discoveries in the Corporate Research and Development Center. Fox's research program developed an improved wire enamel, but in addition discovered an aromatic polycarbonate polymer with extraordinary impact strength. Hay's interest in catalyzed oxidation of aromatic compounds led to a unique polyphenylene oxide polymer, and thence to a moldable Noryl alloy. Sheer good fortune thus played a role in GE's early success, because while neither scientist had targeted a moldable thermoplastic product, these two polymers proved uniquely suitable across a wide spectrum of engineering plastics uses. As Jean Heuschen, GE Plastics Vice-President for World Technology now says, "These are lucky molecules."[14]

After many problem years, GE chemical management had grown in competence and self-confidence from the profitable growth of silicones, and from the diamond success. Lessons learned in silicones and diamonds were successfully applied with Lexan, Noryl, and Valox: notably that product development should take priority over process research, that feedback from the earliest possible sampling of paying customers was critical, that market development and technical service funding were essential from the beginning, and that world position should be sought early.

GE became U.S. and world market leader with Lexan by competitive technology and aggressive marketing, and with Noryl by exclusive technology plus aggressive marketing. By 1987, General Electric had backed these and other engineering thermoplastics with over $3 billion of investment, and having increased these sales to more than $2 billion, had world leadership in engineering plastics. The dollar growth rate of GE engineering plastics from 1968 to 1987 exceeded 25% per year. GE's total of chemical sales in 1987 was $2.75 billion.

CHAPTER 9

Growth By Means of a Major Acquisition: 1988–1991

ABS Plastics Up For Bid; A New Polycarbonate Process

U ntil 1988 GE's engineering plastics business followed a growth-from-within strategy, based on internally generated polymer discoveries and worldwide marketing. Several joint ventures had been formed, notably in Japan and some in the United States, where the association could provide quicker and more effective participation than by handling the project alone. With the Lexan polycarbonates, Noryl modified PPO, Valox polyesters, Ultem-high performance plastic, and important automotive application blends such as Xenoy (PC and PBT) and Noryl GTX (PPO and PA) GE Plastics grew to world leadership in the premium-price, high-performance plastics. Most of these products were technical world-firsts, only the successful Valox line being a me-too technical effort.

Evolution of GE Acquisition and Disposition Philosphy

GE corporate strategy had also emphasized internal growth for many decades, following the turn-of-the-century combinations that created the original General Electric Company. CEO Ralph J. Cordiner wrote in 1956: "General Electric grows from within. It is expanding not by merger or purchase of other companies, but by developing new products and markets – and hence new businesses. "[1]

But as legal, political, and corporate attitudes changed in the 1960s the company developed some growth-by-acquisition initiatives. Utah International, a very successful mining company, was acquired in 1976 for $2.1 billion in GE stock. One rationale for this acquisition was the inflation hedge represented by Utah's mineral reserves and long-term contracts, notably in Australian coking coal. When inflation pressures diminished, John F. Welch (now CEO) decided in 1983 to sell Utah. Ladd Petroleum, a Utah affiliate, was retained at that time, but it was sold in 1990.

Early in his tenure as CEO of GE, Welch stated the strategy of building the company around products and services in which GE was, or could become, No. 1 or No. 2 in the world. "Fix, sell or close" became a prescription for those businesses that were far from meeting those criteria.

Welch expressed the new GE acquisition and disposition strategy in the 1983 Annual Report:

"These dispositions (Utah International, housewares, Family Financial Services) reflect our strategy to focus GE's unique technological, financial and managerial strengths in our 15 key businesses where we believe we can add the most value. ...this strategy has led us to complete 118 additional dispositions totaling more than $1.1 billion over the 1981–1983 period...

"With our increased cash reserves from the sale of Utah and other dispositions, ...the question has been raised: What will we do with the money? The short answer is: It's not going to burn a hole in our pocket.

The cash has given us the flexibility to fund...in 1983 alone, ...62 acquisitions, joint ventures and other equity investments."

GE would now grow by friendly acquisitions if the fit with existing GE businesses was good, so that important synergies could be achieved from the combination, the acquisition cost was appropriate to the expected benefits, and if antitrust regulatory approval could be obtained.

A large acquisition in financial services was completed in 1984 when Employers Reinsurance Corporation became available from Texaco. The cost was $1.3 billion, the largest payout for GE since the Utah acquisition. In 1986, attracted by the leading position of NBC broadcasting and synergies with military electronics, satellites, consumer TV, and semiconductors, GE acquired RCA for $6.4 billion. Some RCA businesses were sold, and the now large United States position of RCA plus GE TV brands was traded in 1987 to (French) Thomson S.A. in exchange for its CGR medical diagnostics (X-ray) business and $1+ billion in cash. Both these large acquisitions had demonstrated a positive value to GE early-on, and although results from a Kidder Peabody securities company acquisition were marginal at this time, GE's corporate leadership had confidence in selective acquisitions as one growth route.

In the 1986 annual report that described the RCA acquisition, Welch also called attention to the plastics business: "GE Plastics, with nearly half its sales outside the United States, has grown 16% annually over the last five years by developing applications in one part of the world, and then multiplying their value through global technology, manufacturing and marketing organizations."

Opportunity for a Large Acquisition of ABS Plastics

Borg Warner Corporation's (BW) Marbon division had added to the field of high-performance thermoplastics in 1954 with the discovery of the ABS resins family derived from acrylonitrile, butadiene, and styrene. As noted in Chapter 8, BW's Cycolac ABS plastics were were less expensive than the nylon polyamides, were easily moldable, and had properties that fitted many applications. BW had chemical sales of $1.25 billion in 1987.

BW's three U.S. polymer plants accounted for almost 50% of the country's ABS capacity, followed by Monsanto and Dow. In Europe BW's ABS share was around 20%, followed by Bayer, Monsanto, and others; in Japan the BW ABS share was about equal among many producers. The BW Chemical Division had also created a U.S. chain of plastic service centers, which were distributors of both high-performance and commodity plastics. BW was the world's largest producer of phosphite chemicals for plastics additives, and it owned half interest in a styrene plant in Mississippi.

BW's other main product was automotive transmissions, and it had some smaller businesses. When BW became the target of two unfriendly takeover efforts in 1986, the company reacted vigorously to thwart these attempts. Purchases of its own stock, sale of a finance company, and, finally, taking the company private by buying out the shareowners in 1987 with the help of Merrill Lynch Capital Partners thwarted GAF Company's unfriendly bid, but left BW with excessive debt. BW rebuffed an early overture from GE Plastics, but subsequently put its chemical business up for sale in order to pay off the large debt that had been assumed.

Borg Warner arranged the negotiation to be a worldwide bidding auction, far different from private bargaining between two parties and their investment advisers. Plastics sales were booming in 1988, so the seller of a successful chemical business could anticipate an attractive valuation. Major chemical and petrochemical corporations around the world were invited to bid, and many did.

The strategic fit of this business with GE Plastics looked attractive. ABS plastics served a large market, with price and properties somewhere between Noryl and the lower-cost, fast-growing polypropylene types. The ABS plas-

115

tics molders and their end-user customers were often the same as for GE's Lexan, Noryl, and Valox. And BW was the world ABS leader, especially in the U.S. BW had also introduced Cycoloy, a useful blend of ABS with polycarbonates, and had begun marketing a Noryl-type plastic (Prevex). GE competitors Bayer and Dow were major producers of ABS. The plastics distributorships were a business that GE Plastics had considered and wanted to enter. Phosphites additives and a styrene monomer facility were interesting and related assets. Acquisition would increase GE's Plastics' sales about 45% and operating profits about 37%, based on 1987 financial data.

On the downside, ABS technology was becoming mature and was being practiced by more world competitors than GE was facing in its other plastics. BW's early technical leadership no longer prevailed with some customers. GE's major appliance business, for example, favored Monsanto as a large ABS supplier for refrigerator parts (this $50 million ABS potential internal to GE was a plus opportunity). ABS markets were more like the commodity plastics, with larger tonnages, more competitors, narrower margins, and more price competition. GE would be bidding for a successful organization, but one in which the personnel had previously viewed GE Noryl products as competition. Would GE's market development skills enhance an ABS product line, or would culture differences prevent an effective combination?

And finally, would the bidding process get to a level at which a satisfactory return on the investment could not be earned? Clearly there is some price level where an acquisition is no longer attractive. Some of the competing bidders would be considering synergies similar to those for GE. Petrochemical producers of one or more of the three monomers would be bidding to integrate forward, perhaps to enter premium plastics production for the first time. BW management hoped that GE would be the successful bidder. GE wanted the acquisition, and also preferred not to see a strong petrochemical competitor buy a niche in engineering plastics.

The GE management team considering the ABS acquisition were Glen H. Hiner, Senior Vice-President, GE Plastics; Paul L. Dawson, CEO for GE Plastics B.V. (Europe); Edward R. Koscher, Vice-President Sales (US); Herbert G. Rammrath, President, GE Plastics Pacific; L. Donald Simpson, Vice-President Manufacturing (U.S.); Uwe S. Wascher, Vice-President, Marketing; and Joseph G. Wirth, Vice-President, Technology. Hiner reported to corporate Vice-Chairman Lawrence A. Bossidy, and the latter to CEO Welch. As the bidding approached $2 billion, Hiner's team is reported to have had a majority in favor, but two negative votes. Hiner was strongly in favor, as was Bossidy. Welch was positive. European and Japanese bidders finally withdrew, leaving GE and Exxon; and GE was the successful bidder at a price of $2.3 billion in cash. The agreement with Borg Warner was announced in June 1988, and was consummated in the third quarter after regulatory approvals.[2,3] During this acquisition year, the assets associated with GE Plastics jumped from $3.9 to $7.1 billion, an increase of $3.2 billion. This acquisition was a *big* deal.

GE estimated that the sales gained as of 1988 were about $1.6 billion. Thus the company was paying about $1.44 per dollar of sales acquired (2.3/1.6). The bidding had brought this to a high ratio, though not out of line with GE Plastic's end of 1987 ratio of assets to sales, $1.42 (3901/2751). However, the Borg Warner product lines would not produce as good margins on sales, and therefore could not be expected to earn as high a return on the large new investment.

Integrating the Acquisition

GE Plastics' (GEP) plan to integrate the acquisition was carried out promptly. The ABS resins and related products were merged into GEP's U.S. operations and the two resin sales forces were combined. Three separate profit and loss (P&L) business units were created that reported to the head of GE Plastics: (1) the plastics distributorships, renamed Polymerland; (2) phosphites and other additives, to be called Specialty Chemicals; and (3) petrochemicals, consisting of the 50% interest in a Louisiana styrene plant and a one-sixth interest in an ethylene supplier. The attractive GE benefits were phased in for the new employees.

Acquisition in Hindsight

Some expected acquisition synergies were realized, others were not. The Cycoloy blends of ABS and polycarbonate proved important. The acquired Prevex products and customers added to GE's Noryl marketplace. Integration of people and plants into GE Plastics went according to plan and job assignments were fairly made, but a large number of Borg Warner skilled sales reps left the company and some executive placements did not become long term. The Polymerland distributorships grew rapidly. The styrene plant joint venture has proved profitable, particularly during the business cycle upside.

Competition in ABS plastics has increased. Improved polypropylene grades have begun pushing into applications formerly served by ABS. Also, while GE is still the ABS market leader in the U.S., price competition in the simplest technology grades has produced a new world volume leader, the Taiwan-based Chei-Mei Company. Asian-based ABS capacity substantially exceeds demand. Monsanto offered its $700 million styrenics (ABS) business for sale in 1995 and the purchaser at year-end was Bayer. Perhaps reflecting the changing ABS competitive situation, the purchase price was reported to be $580 million, or $0.83 per dollar of sales (a ratio 40% less than GE had paid seven years earlier for its ABS acquisition).[5]

Gary Rogers, the current President and CEO of GE Plastics, notes that the acquisition strategic fit was very good, but that it was expensive.[6] Welch now says that GE paid too much, maybe by half a billion dollars. He suggests that the purchase will not be regarded as one of the best during his leadership.[7]

Mixed conclusions may be drawn from GE Plastics' overall financial results following the 1988 acquisition year. World plastics demand remained strong for 2 years, then declined seriously, and finally recovered in 1994-1995. As shown by Figure 9-1, following the acquisition GE Plastics sales and profit jumped in 1989 and 1990, and the profit ratios held steady. As demand declined in 1991–1993, sales and profit dropped well below the 1989–1990 levels, as did the ratios of operating profit to sales and to assets. In 1992 these ratios reached a very unsatisfactory low of 15% and 9%. Recovery from 1994–1997 restored sales, operating profit, and the two profit ratios, all to record levels. These annual report data do not separately disclose the acquired products results per se, nor their synergies. The acquisition certainly gave GE Plastics a much larger position in the world plastics market, and despite narrow margins in ABS products per se, the overall returns on sales and assets after one business cycle are higher than before the acquisition.

Polycarbonate Production in Japan with a New Process[8]

Hiner's negotiations in Japan in 1989 rearranged the corporate structure of the GE Plastics joint ventures there. Engineering Products Ltd. (EPL), the 1971 marketing and compounding joint venture with Nagase, was combined with the GE-Mitsui ventures in polyphenylene (PPO) and bisphenol A (BPA) to form GE Plastics Japan, Ltd., in which GE Plastics has a 51% ownership.

Mitsui and GE personnel were already planning a new polycarbonates plant for Japan. A critical issue was whether to install the proven interfacial polymerization of BPA in caustic and solvent with phosgene, or to attempt development of the "melt process," reacting BPA and diphenyl carbonate. If the melt process was chosen, then a synthesis for diphenyl carbonate was needed, either by the reaction of phenol with phosgene, or by some other route. GE Plastics was then operating three world-scale plants with the continuous interfacial process: Mt. Vernon, Indiana, Burkville, Alabama, and

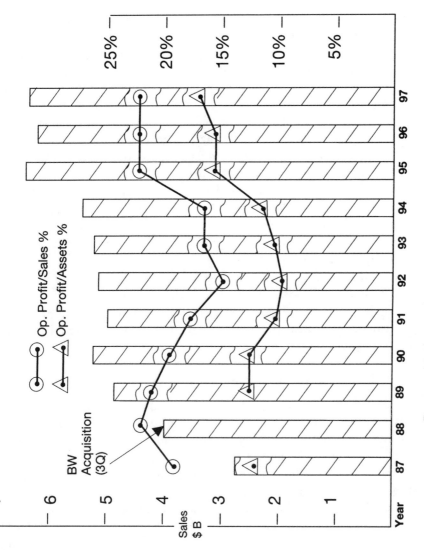

Fig. 9-1. GE Plastics sales and profit ratios 1987-1997.

Bergen op Zoom, The Netherlands. All competitors were believed to be using similar chemistry, but during the years since his 1953 discovery, Daniel W. Fox had continually championed the melt process, which would eliminate methylene chloride solvent and potentially offered lower cost. Some pilot-plant demonstrations had been made at Mt. Vernon by Robert Edwards and Fred Bunger in 1977–1979. GE technical managers were deeply divided on the process choice issue. Mitsui leaders favored developing the melt process, and this high-risk alternate was finally selected. Pilot-plant work at the Mitsui Petrochemical Iwakuni facility was very successful and the melt process was installed at Chiba, Japan with a 55 million pounds per year plant (see Fig. 6-1).

The reaction of phenol and phosgene for making diphenyl carbonate was well-known, but the idea of eliminating phosgene from the flow sheet was an attractive possibility. At one point a Japan appropriation request, including a phosgene reaction, was turned down by GE CEO Welch, who challenged the joint venture partners to develop some diphenyl carbonate technology purchased from Italian Enie-Chem Company. This was successful in time to install a carbon monoxide–methanol–phenol process at the new Chiba facility. The two-step process for diphenyl carbonate is summarized in these equations (Fig. 9-2):

The Chiba, Japan, plant thus became first world production unit to make polycarbonate plastic without either methylene chloride solvent or phosgene. This polycarbonate resin would not see a chlorine or a sodium atom. GE Plastics management predicted that the new flow sheet would further reduce the cost of making its flagship plastic, Lexan.

GE Plastics in the Pittsfield Community

While GE Plastics was growing rapidly and adding to its Pittsfield, Massachusetts headquarters in the 1980s, GE's large transformer business had come upon hard times. A deep drop in demand from U.S. electric utilities, combined with stiff import competition, diminished the prospects for giant transformers produced in the sprawling Pittsfield Works. This plant's history went back to GE's acquisition of the Stanley Manufacturing Company in 1906. The Pittsfield Works and its laboratory had spawned the phenolic resins and compounds, the phenolic laminates, and the molded parts businesses, products that made up the largest part of GE's Chemical Division at its 1945 formation and for several years thereafter.

In 1986 GE announced withdrawal from the power transformer business and shutdown plans for the large facility. As transformer manufacture wound down, the modern GE Plastics acquired and converted the largest

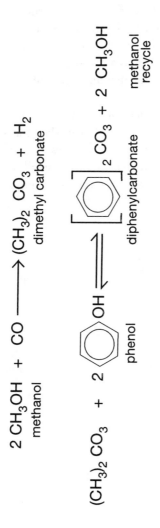

$$2\ CH_3OH\ +\ CO\ \longrightarrow\ (CH_3)_2\ CO_3\ +\ H_2$$
methanol   dimethyl carbonate

$(CH_3)_2\ CO_3\ +\ 2$ [phenol] $\rightleftharpoons$ [diphenylcarbonate]$_2$ $CO_3\ +\ 2\ CH_3OH$

methanol recycle

Fig. 9-2.

high bay transformer building into a Polymer Processing Development Center, equipping it with many kinds of plastics-molding and extrusion equipment, including a giant Alpha 1 press imported from Germany. The most recent equipment addition replicates the process by which customers mold special polycarbonate plastics to form CD's.

The Living Environments House in Pittsfield is a GE Plastics application demonstration. This 3000-sq.-ft. residence structure opened in 1989 after several years of research and study. In it, present and future applications of many plastics and composites with concepts drawn from every corner of the world, are displayed. It serves as a test site for products like plastic shingles, siding, and glazing, and for the demonstration of many design ideas in structure, fittings, and appliances.

In 1992 GE's Ordnance plant in Pittsfield, designer and builder of high-tech military hardware, was sold to Martin-Marietta as part of a large divestiture of GE's military and aerospace business. After this sale, GE Plastics, which for more than 45 years had been the smallest General Electric organization in the community, was now the only company business in town.

CHAPTER 10

Laminates and Insulating Materials
GE Core-Businesses
Decline in Importance

G E Electromaterials, with headquarters and plant at Coshocton, Ohio, had its early product genesis in phenolic resin-paper laminates for electrical insulation. When the GE Chemical Division was formed in 1945 (Chapter 2), these laminates were manufactured in the Lynn, Massachusetts, Works; the phenolic resins were produced within the Pittsfield, Massachusetts Works; and product development flowed from a substantial Plastics Laboratory on the Pittsfield outskirts at One Plastics Avenue. Laminates were the second largest product line of the newly formed division (after plastic molded parts).

In 1948 all laminates production was moved to a new plant in Coshocton, where phenolic resin kettles had been installed. This move was caused by GE's new aircraft jet-engine business's space needs, and by the need to increase production of inner doors for GE refrigerators and those of other appliance manufacturers. Competitive production of industrial laminates was well-established at that time in companies such as Continental Diamond Fibre, the Micarta division of Westinghouse, Formica Corporation, Spalding, and Pioneer.

Refrigerator Strips and Inner Doors

White laminate separator strips for GE refrigerators had been developed in 1940, and the immediate post–World War II refrigerator designs also used a white-painted laminate inner door, so as to stretch the limited allocation of sheet steel. A patent covering shelves in refrigerator doors was at that time exclusively held by the Crosley Corporation (trade-named Shelvador). Refrigerator inner doors amounted to half of the entire Coshocton plant output when the Crosley patent expired in 1952. At that time all refrigerator manufacturers rushed to offer models with shelves in the door. The preferred technology for these inner-door panels would be vacuum-formed polystyrene sheet, a convenient process using inexpensive molds and well-suited to in-line production in a refrigerator factory. Facing the certain loss of half the business over the next few years, GE's Laminated Products organization focused on two initiatives: (1) decorative laminates; and (2) expansion of industrial laminates applications, particularly in electronics manufacture.

Decorative Laminates: GE Textolite

GE's decorative Textolite had been withdrawn from the consumer market after war broke out in 1941, though some orders from military base construction were filled. Postwar, the demands of starting up the new Coshocton plant and of increasing production for refrigerator inner doors again deferred the company's capacity to market the decorative product line effectively.

Formica Corporation, later a division of the American Cyanamid Company, had pioneered the market development of decorative laminates, and had built a large new plant outside Cincinnati, Ohio especially for the product. Presses in this Formica plant were sized to produce sheets that were nearly double the area of those produced by GE's largest industrial laminate presses. With a new plant ready to produce soon after the war's end, Formica signed up the best construction materials distributors in major market areas and became the unchallenged market-share leader. With colorful designs, early architectural contacts, and effective distribution, "Formica" became a generic term for the popular product.

Decorative laminates for countertops were a great improvement over linoleum surfaces, and the market boomed. Furniture makers brought out lines of tables surfaced with this almost indestructible material. Marketing and distribution channels for decorative laminates were completely different from industrial laminates, requiring different skills and organization. But the decorative market's rapid growth in the 1950s and its large size compared to industrial laminates encouraged GE and other competitors to chase the market leader.

Some early sales volume and market-share gains by GE encouraged the

124

management to continue the effort, though the profit margins in the decorative laminate line never approached those of the also growing industrial line. GE was a leader in industrial laminates, but was a poor third in decorative products, behind Formica and Wilson-Art. A long succession of GE managers at Coshocton continued to push the decorative product as well as the industrial. To increase capacity for both lines, GE bought Parkwood Laminates (in Massachusetts) in 1974. This decision proved very unfortunate, especially as the entire laminates market plummeted in the recession the following year. Parkwood was resold and GE later exited from decorative laminates, selling the business to Maryland-based Nevamar, Inc., in 1979.

Leadership in Industrial Laminates

GE was well-positioned in laminates for electrical uses, having been in the business since the 1920s and having a large volume of internal sales for electrical equipment. From the 1950s on the largest growth and technology change for industrial laminates came from the ever-changing and explosively growing electronics manufactures: radios, TV, industrial controls, and computers. Large applications would warrant a National Electrical Manufacturers Association (NEMA) standard specification; other grades were developed for the special needs of large customers. The GE organization became skilled in serving the changing needs of the market and particular customers.

Electronics packaging of the 1950s and 1960s shifted to the printed circuit board, faced on one or both sides with a thin film of copper. The circuit design interconnections would be printed on the copper, which was then etched away chemically except where the desired connections had been printed. Holes punched in the board accepted leads from resistors, capacitors, transistors, and integrated circuits, and the backside of the board would then be dipped into a molten solder bath to make the desired connections. For this application GE and other producers developed copper-faced, epoxy-bonded glass laminates that could withstand the chemical attack as well as the soldering process temperature.

Computer industry growth was a big factor in electronics production in the 1960s and 1970s, with IBM's needs by far the largest in the industry. GE met this customer's specs so well that IBM became the largest single customer of the Coshocton plant. GE thus managed to replace the 1950s loss of refrigerator business with a large number of new applications, though the IBM account was subsequently lost when the customer decided to make its own circuit boards. The circuit-board market continued, but subject to major technology change. Because integrated circuits on a single chip now replace thousands of the former circuit boards, the market size has shrunk many fold at the laminator level.

GE Electromaterials remains a market leader in electrical laminates, though the business today is small compared to the company's engineering plastics and silicones product lines. The present product lines include "FR-4" copper-clad epoxy-glass laminates, in rigid, thin and formable sheets, as well as "CEM" cold-punchable single-side types. A "GETEK" product line of polyphenylene oxide/epoxy-glass laminates are designed for specially high electrical performance. These GETEK products are also made and distributed by Matsushita Electrical Works, Ltd., under license.

Other Insulation Products: Insulating Materials Department

As noted in Chapter 2, other electrical insulations being manufactured by GE early on were mica products, electrical impregnating varnishes, insulating tapes of many kinds, wire enamels, and certain industrial paints. All were originally produced in scattered buildings around the Schenectady, New York, Works. In 1949 the Resins and Insulating Materials Department (RIM) product responsibility was divided between the Laminates Department management (the solid products) and that of the Chemical Materials Department (the liquid products).

After GE's exit from alkyd resin manufacture (1956), division manager Robert L. Gibson decided to see if increased management focus on insulation products would improve results and increase external sales. First as a product section within Chemical Materials, and then in 1959 as the separate Insulating Materials Department (IMD), a management team focused on improved products, distributor development, and selected original-equipment-manufacture (OEM) direct-sale customers.

Alkanex wire enamel from the R&D Center project (Chapter 6), proved as great a success in electrical machinery as had the Formex product 20 years earlier. And a follow-on enamel, Isonel, extended temperature performance further. But GE's patent position in these products did not give the company a decisive competitive advantage. Independents Schenectady Chemicals and P. D. George companies were strong competitors in wire enamels and insulating varnishes.

Mica flakes, a mined material, make extraordinarily effective electric insulation in respect to dielectric strength. In tape form, or as molded shapes, the mica flakes are bonded with varnishes to seal the interstices and eliminate air tracks for electrical arc-through. One unique mica process innovation enhanced the GE tape products. "Micamat" was a development of the

Plastics Laboratory around 1949, in which a slurry of small mica flakes was dewatered through a fine wire mesh, in the same way that paper is formed from a pulp slurry. After heat drying, the "mica paper" held its sheet form and could then be treated with various varnishes and backings to form flexible laminates, and used as sheets or cut into tapes. The big technical leap from a small fixed-screen laboratory demonstration to an actual paper machine with a continuously moving Fourdrinier wire screen was demonstrated in Pittsfield and then moved to the Coshocton plant, where the Micamat process runs essentially unchanged today. Because small flake mica costs less than large flakes, and because the mica paper machine is less expensive than machine-laid flake processes, Micamat tapes became preferred and have largely replaced the large-flake product.

The newly formed IMD moved its headquarters and most manufacturing processes to Schenectady's outskirts, a location called Riverview. Industrial paints manufacture had been moved to a plant in Chelsea, Massachusetts, after the 1956 plant explosion. As with other division products, IMD had to earn, and compete to hold, the insulation business from GE electrical manufacturing departments. The reorganization ordered by Ralph J. Cordiner had decreed that internal sales transfers be priced at external market levels, and that a purchasing department was free to buy from outside sources if better quality or value were offered. IMD often found it difficult to prove product-quality equivalence if competitive materials ever became entrenched. Another source of internal tension came when a cooperative internal technical development resulted in a novel insulating material. Would the buying department be entitled to exclusive use, and thus a competitive advantage? The IMD position: a reasonable lead time would be granted for exclusive use, but this would be followed by broad marketing of the new product. This policy was often a tough sell with GE electrical departments.

Externally, distributors of insulating materials welcomed quality insulations from GE, but rarely offered these exclusively. In each of its product lines GE's IMD organization faced several competitors, so the good distributors could pick and choose. External direct-sale OEM customers were interested in new developments, but they were understandably reluctant to buy from an arm of GE, their electrical competitor.

The IMD sales in 1957 were around $12 million, of which less than 20% were external. At its peak in the 1970s, the sales reached around $40 million, and external sales had grown to 35% of the total. In addition to the wire enamels, Permafil-type impregnating varnishes (Chapter 4) were a GE leadership product. Another growth product was cable accessories, a kit for terminations and splicing. Customers for cable accessory kits were mainly electric utilities, and this product line was eventually transferred to an accessories arm of GE's transformer business. The small "Glyptal" paint line was sold in the 1980s.

With diminished sales, the organization status of IMD within GE was reduced to a product section, first within the Laminates Department and later under Silicones. While the assemblage of insulating products remained profitable, it was small and showed little growth. Adequate competition existed to serve the GE electrical product plants, so there was no compelling company reason to continue the organization. The facilities at Riverview and all the remaining insulation products were sold to a private investor in 1988. That party resold the business in 1995 to a Swiss-owned company, Von Roll Isola USA Inc.

Thus, the laminates and insulating products, which were important contributors to sales and earnings in the early years of the Chemical and Metallurgical Division, diminished greatly in relative importance in more recent times. But we should remember that GE's search for better electrical insulation produced the 1940s discoveries in silicones; and if GE had not sought a better wire enamel in 1953, Daniel W. Fox would not have discovered Lexan.

CHAPTER 11

GE Silicones:
1965–1998

Sealants Leadership;
World Participation

In Chapter 3 we traced the growth of GE Silicones from plant opening in 1947, through difficult years, and finally to successful business performance in the 1960s. GE silicone sales in 1965 were $32 million and were growing rapidly. In 1968 they were around $42 million, at that time substantially greater than the combined GE engineering plastics, Lexan and Noryl. Also by this time GE had finally built a separate office building at the Waterford, New York, site. In prior years laboratories and plant capacity additions held top priority for capital investment allocation. Dow Corning (DC) silicone sales in 1968 were around $105 million.[1]

Sealants and Other RTV Rubbers

A large part of GE Silicone's sales growth in the decades since 1965 has come from the broad product class of room-temperature vulcanizing (RTV) rubbers. GE initially introduced these flowable rubbers that cured at room temperature in 1959, as two-part formulations. GE RTV 60 and several other two-part systems were soon accepted as sealants for commercial airframes. Market development for long-lasting construction sealants of many types followed, as GE developed the technical performance data to support promotion to architectural specifiers. The construction industry welcomed these new products, because silicone rubber's long-term resistance to sunlight and ozone exposure is unique among elastomers, and both GE and DC offered a long-life sealant material guarantee. GE also moved early to establish effective distribution to the construction industries.

The development of one-part RTV formulations by all competitors made it possible to use caulking cartridges and small toothpaste-like tubes for consumer use. GE marketed the construction sealants in several colors. To market the initial five color choices of the consumer tubes, GE employed the company's lamp distribution channels to food and hardware outlets.

The early silicone RTVs were formulated from flowable viscosity gum polymers containing silanol (Si-OH) end groups. Two-part systems used organic tin compounds to catalyze the silanol condensation after mixing, while in one-part systems the condensation and cross-linking were activated when atmospheric moisture reacted with a silicon-acetoxy or a silicon-alkoxy molecule. An additional curing system was introduced in flowable silicone rubbers in recent years. If a two-part RTV system composed of a vinyl-mod-

Fig. 11-1. GE Silicones RTV rubber sealants are available in a broad line of caulking cartridges.

130

ified dimethylsiloxane gum containing a platinum-compound catalyst is mixed with a dimethylsiloxane gum containing Si-H units, a cross-linking is effected at room temperature with no by products. This reaction, originally discovered by Dow Corning's John Speier,[1] is important not only in RTV-type applications but also in conventional rubber fabrication and with some special resins.

RTV applications as different as mold-making for casting liquid plastics, electronics encapsulation, and even dental impressions stimulated continual expansion of the product line and its packaging. Successful marketing of the broad range of RTV products supported a significant share of GE silicone sales and earnings growth in the 1960s, 1970s and 1980s. Construction sealants for new buildings, window glazing, and consumer use proved a very large market, and GE has held a leadership position in these applications. As RTV sales grew, GE developed in-plant process improvements for continuous gum polymerization, compounding, and packaging that dramatically increased manufacturing productivity.

During the period when Thomas H. Fitzgerald was General Manager (1972–1984), GE's success with the RTV products was extended to Europe, as the company began a direct selling effort there. GE also began manufacturing on the continent in the 1970s with a compounding and packaging plant at Bergen Op Zoom, The Netherlands, on the site already developed by GE's engineering plastics.

GE-Toshiba Silicones

In the 1960s GE silicone exports to Japan and Asia were small compared to sales of the leading local competitors, Shin Etsu and the Toray–Dow Corning joint venture. Toshiba Silicones, a GE licensee, was a small factor at the time. GE Silicones general manager, Robert T. Daily, discussed, but was not able to conclude, a joint venture proposal with Shin Etsu. He did reach agreement on a 49% GE joint venture with Toshiba in 1966, which was consummated in 1971. After a faltering beginning, GE–Toshiba Silicones, or Tosil, became an effective competitor. One key to success, initiated by Joseph G. Wirth, then GE Silicones R&D manager, was a regular technical personnel interchange, which, over time, greatly benefited both organizations. A change in U.S. joint venture accounting practices encouraged GE to negotiate a 51% position in Tosil in 1994. This allowed GE to report the total sales and operating profit on its U.S. books, along with appropriate elimination of the partner share of profit when calculating GE net earnings. Tosil today claims a strong number-three position among Japan silicone producers.

Other Silicone Rubber Markets

Other silicone rubber markets have grown explosively as new technology improved the curing systems, fabricator processing costs dropped, flowable rubber compounds were developed, and overall prices were reduced. Cured silicone rubber is unique among competing elastomers for long life at high temperatures; resiliency at low temperature; for sunlight, ozone, and air-oxidation resistance; and for excellent (low) compression-set characteristics.

Until recently, silicone rubber compounds containing benzoyl peroxide were fabricated by molding or extrusion and heat-vulcanized by oxidation of methyl or vinyl groups. Another curing chemistry, the platinum-catalyzed two-part system, described earlier for RTVs, is also employed for fabricated parts. GE's line of such rubber compounds is trade-named LIM (liquid injection molded), which can be mixed with a catalyst and pump-fed en route to a heated mold. When high parts volume justifies the expensive mold cost, fabrication by this method can be less costly than if the parts were molded with conventional heat-cured silicone rubber.

Whereas aircraft applications were a large market factor in the 1950s and early 1960s, silicone rubber parts are now also important in automotive engines and many other mechanical designs. Molded-in-place gaskets are possible in some high-volume automotive applications. Some computer keyboards get their resilience from a molded silicone rubber pad. Baby-bottle nipples and some children's toys are now made from long-lasting silicone. Silicone rubber wire and cable insulation have also expanded beyond military applications to many commercial uses where high temperature or oxidation resistance is a factor.

Silicone rubber competitive strategies have changed over the years. In the 1960s and 1970s, both GE and DC marketed a broad range of ready-to-fabricate compounds, but also offered reinforced-gum formulations to both fabricators and custom compounders, the latter serving fabricators and wire and cable makers. GE added a custom-compounding location in California, while DC made custom compounds at several locations apart from its main plant. In 1983 Fitzgerald initiated the development, led by Joseph C. Caprino, for a GE "Sil-Plus" line of reinforced-gum bases designed especially for the custom-compounder trade level. GE then reduced marketing emphasis on custom compounds produced at the Waterford plant, except for automotive applications.

Wacker, a German producer, became a significant factor in U.S. silicone rubber markets in the 1970s by purchasing a half-interest in Stauffer's silicone business, and then in 1987 increasing its ownership to 100%. Wacker's plant at Adrian, Michigan, imports siloxanes from Germany, and offers reinforced gums, RTV polymers, and fluids.

132

Fig. 11-2. GE silicone elastomers improve performance and extend service life in products as diverse as electronic keyboards (above) and baby bottle nipples (below).

Fig. 11-3. Baby bottle nipples

Fluids

As the number of U.S. competitors increased from two to four in the 1960s, prices of conventional fluids declined. Additional competition from European producers accelerated the price decline when the basic U.S. fluids patent (Patnode) expired in 1966. A partial price recovery occurred in the 1970s.[2] Traditional fluids application in polishes, mold release, textile finishing, grease manufacture, antifoams, and plastics additives have continued to grow.

One of the largest silicone fluids growth markets of recent years has been in personal-care products, such as underarm antiperspirants and shampoos with hair conditioners included. The silicone products used include conventional dimethylsiloxane fluids, high-viscosity fluids, very-high-molecular-weight gum polymers, controlled volatility silicone liquids, and some siloxane–organic block copolymers.

World expansion of urethane foam rubbers and rigid foams has carried with it growth of polyurethane foam additives (PUFA) made from silicone–polyether block copolymers. Witco, which purchased the Union Carbide silicone spinoff, is a leader in this application. GE had products in the PUFA market for a decade, but withdrew in 1974, not able to compete effectively with Union Carbide's complete market basket of urethane foam ingredients.

Fig. 11-4. GE silicone fluids enhance many shampoos and hair conditioners.

Silicone Resins and Specialties

Conventional silicone resins continue to be used in high-temperature and weather-resistant paints, electrical impregnating and mica- or glass-bonding varnishes, and as the bonding agent for silicone/glass-cloth laminates. In recent years GE has pioneered a silicone resin technology known as "Hardcoat," which is based on monofunctional trimethylsiloxane "M" units and quatrofunctional siloxane "Q" units. This resin is applied as a transparent overcoat to many plastics, especially to polycarbonate headlight lenses for improved abrasion resistance.

The growing markets for (1) pressure-sensitive adhesives (PSA), and (2) release-paper backings for sticky tapes have created large silicone specialty opportunities in both applications. GE offers a line of sticky PSA materials for labelmakers, especially where a wide temperature-range use requirement is best met by silicone formulations. Recent large applications for silicone-treated release papers are the stick-on note pads and the backing paper for new peel-away U.S. postage stamps. Some resin systems in this GE line, developed by Frank Modic, use the platinum-catalyzed Si-vinyl plus Si-hydrogen additions. Similar silicone gels are now used to coat the fabrics in automotive airbags, and also to create a new type of roof coating for buildings that is spread over polyurethane foam.

Chlorosilane Intermediates

Silicone fluids and rubber markets each exceed the volume of silicone resins. Since the polymer chain of both fluids and rubber consists chiefly of dimethylsiloxane "D" units, it therefore follows that a basic silicone producer must operate the direct process efficiently to maximize the output of the D-unit precursor, dimethyldichlorosilane. Process research by GE, and no doubt by all the successful basic producers, has greatly increased the conversion efficiency of methyl chloride and silicon to dimethyldichlorosilane, and the reaction is carried out in ever-larger fluidized-bed reactors in a semi-continuous mode. GE has added several fluidized bed reactors and extensive distillation systems to its Waterford plant. World silicon production has grown with the silicone industry until this usage has become the largest of any elemental silicon application. Elchem (Norway) is the major world supplier of silicon.

Trichlorosilane ($HSiCl_3$), another major intermediate, is made from the high-temperature reaction of hydrogen chloride and silicon metal. It can then be converted to phenyltrichlorosilane and other organochlorosilanes. Trichlorosilane can also be burned in oxygen to form finely divided "fumed silica," which is used to reinforce many silicone rubber compounds. GE

135

integrated backward in the 1980s with a fumed silica production unit, and also purchased some of the filler from a neighboring unit of Degussa Company, the original technology developer.

Purified trichlorosilane is also the feedstock for processes producing hyperpure silicon, the precursor to the single-crystal silicon wafers needed to make memory and integrated-circuit chips. In 1959 Dow Corning entered the field, and has remained a producer of hyperpure silicon.[1] GE Silicones purchased the Great Western Silicon Company (Arizona) in 1980, but this venture was not successful and was resold to a Japanese company, NKK, in 1987.

From 1954 on GE made phenyltrichlorosilane and diphenyldichlorosilane by reacting silicon, using copper catalyst, with a mixture of chlorobenzene and hydrogen chloride at high temperatures (ca. 550°C). Because some polychlorinated biphenyls (PCBs) are a byproduct of this reaction, the environmental control and appropriate waste disposal was a continuing concern. In 1985 GE shut down the process and chose to purchase phenylchlorosilane intermediates from other suppliers.

Waste Disposal

GE, along with all chemical manufacturers, has responded to the greatly tightened federal regulations for emissions and waste disposal by continually adding state-of-the-art control and disposition processes. These now include bacterial conversion and separation of organic materials from wastewater; two different incineration processes, each with a flue-gas afterscrubber; and final neutralization of residual wastewater acidity. These waste process additions made up a significant proportion of capital investment at the GE site in the recent decades.

Competition and Industry Structure

Over the years the world silicone industry has undergone considerable restructuring. While GE in the United States offers a broad line of fluids, rubber, resin, and specialties, its particular emphasis has been fluids and rubber of all types, and especially the many RTV sealants applications. GE has withdrawn from making some specialty rubbers, such as dental-impression RTVs, and from grease compounding. In addition to the basic silicone producers, meaning those that operate the direct methylchlorosilane process, a host of specialist competitors operate at various levels. Compounders of sealants and specialties can purchase dimethylsiloxane gums and fluids containing Si-OH (silanol), Si-vinyl (vinylsiloxane), and Si-H (silane modified) reactive sites from several competitors.

Of the early silicone producers, ICI (UK) was unsuccessful and was first to exit from the industry, selling some assets to Rhone-Poulenc.

The Union Carbide (UC) Silicones division inititially managed a broadline basic operation, but later concentrated on specialty positions, most importantly the PUFAs. UC Silicones first withdrew from making silicone rubber and then joined organizationally with a larger UC chemicals division, using a matrix management approach. When a new methylchlorosilane facility, built in 1987 at the UC multiproduct South Charleston, West Virginia facility was shut down, UC purchased siloxane intermediates rather than operate the direct process.

In 1987 GE and UC announced agreement to merge their silicone interests in a joint company. Although the two product lines at that time had limited overlap, the Federal Trade Commission objected to the proposal. The two companies also had second thoughts and agreed not to pursue the venture. UC proceeded with its corporate plan to divest itself of silicones, first spinning the business off to private investors. This O-Si Specialties Company was acquired by Witco in 1995.

We have noted Wacker's two-step acquisition of Stauffer Chemical's silicone position to serve as its U.S. operations base, and also the GE-Toshiba successful venture in Japan.

GE had recent plans for building a basic silicone facility at Bergen op Zoom (BOZ), but instead opted for a European joint venture. In December 1997, GE Plastics and Bayer AG of Germany announced plans to merge their silicone interests in Europe, Africa, the Middle East, and India. Headquarters and the main plant of this company will be at Leverkusen, Germany, where basic capacity will be expanded. GE silicone facilities at BOZ, and in England and India are included in the package, as is a Bayer plant in Italy. News announcements stated that the combined sales would be about $410 million, with GE owning 50.1% of the company. This merger was consummated in July 1998; so GE, at long last, has a basic silicone manufacturing presence in each of the three world industrial areas.

World Silicone Industry

World silicone markets have been stimulated by the extraordinary variety and versatility of many silicone polymers as well as by price reductions. Market growth over the past three decades has averaged around 12% per year in dollars and somewhat more in volume. Some current prices (1998) are roughly half those prevailing in 1965.

GE estimates world silicone volume for 1997 at $5.7 billion. Dow

Corning has maintained a wide lead among silicone competitors in most areas of the world, with most recent sales reported at $2.6 billion. GE Silicones consolidated sales for 1996 were reported to be $1.03 billion.[3] Interestingly, today's ratio of GE to Dow Corning silicone sales is essentially the same as existed in 1965.

Estimated 1998 area rankings of the larger competitors are noted below: (B) indicates the basic methylchlorosilane manufacturers in the area.

U.S. & Americas	Europe & Middle East	Japan & Asia
Dow-Corning (B)	Dow-Corning (B	Shin Etsu (B)
GE (B)	GE-Bayer (B)	Toray-Dow Corning (B)
Wacker	Wacker (B)	Tosil (GE) (B)
Witco	Rhodia (B)	Rhodia
Rhodia	Huls (B)	
Shin Etsu	Goldschmidt	

CHAPTER 12

GE Engineering Plastics: 1992–1998

After Recession, Growth Resumes

Leadership of GE Plastics (GEP) shifted at year-end 1991 to Gary L. Rogers, who transferred from heading GE Appliances when Glen H. Hiner resigned to become Chief Executive of Owens Corning Corporation. In the 1991 GE annual report Rogers announced the 1990s agenda for GEP: to be the world's highest-quality, lowest-cost producer of engineering materials.

The world recession that began in 1990 bottomed for plastics in 1992. GEP management reacted to the slowdown with a wrenching $50 million expense reduction–including more than 200 salaried people laid off in Pittsfield, Massachusetts, the first major personnel reduction in the business's history–by a cutback in the rate of new plant additions, and by the sale of some assets. But as the underlying growth of markets and products continued, and economic activity finally turned around, GEP sales and earnings rebounded strongly through a combination of volume increase, productivity gains, product-line evolution, capacity expansion, organization restructuring, and a powerful new management tool known as 6 Sigma.[1]

Polycarbonate Capacity Expansion

Programs to stretch the capacity of existing polycarbonate process lines at Mt. Vernon, Indiana, and Bergen op Zoom (BOZ) added 150 million pounds in 1993, and the large-scale unit at Chiba, Japan, added an equal amount. Additional process lines at Burkville, Alabama, added 110 million pounds in 1996. Rogers noted that when the 285 million pounds of new process (non-chlorine) Lexan capacity at Cartagena, Spain, came on stream in 1999, GE had doubled its world PC capacity since 1993, and that production costs would be half of that year's level.[1,2] In July 1998, GE Plastics announced that a second polycarbonate process line of equal capacity would be installed at Cartegena, to come on-stream in 2002.

Polycarbonates were GE's first thermoplastic, and they remain the most important. A recent *Chemical and Engineering News* world review estimates GE's share of the world's 3.2 billion lb/yr PC capacity at 43%, followed by Bayer at 29%, Dow at 12%, Teijin at 7%, Mitsubishi Gas Chemical at 3%, and others totaling 6%. A major polycarbonate growth driver now is the application in compact discs (CDs) and the newer digital video discs (DVDs). Ever more advanced data storage media are under development.[2]

Polymerland

Another fast-growing sales unit in recent years has been Polymerland, the plastics distribution business acquired with the Borg Warner acquisition. This distribution network focuses on smaller molders and sells not only GE engineering plastics but also the polyamides and polyacetals of other producers, plus a broad line of commodity plastics: polystyrene, polypropylene, selected polyethylene, acrylics, and polyvinyl chloride. About 40% of Polymerland's business (by volume) is now represented by GEP products and 60% by other producers.[3]

GEP began expanding this distribution in the United States by opening four new locations and acquiring two other distributors. Expansion in Europe came next, with outlets added by acquisition and new building, in Germany, France, the U.K., Benelux, and Spain. A Polymerland joint venture was established with TIPCO in India in 1995, and in 1996 distributors in Turkey and Scandanavia were acquired. Current Polymerland sales volume in the United States is estimated at $750 million, with an additional $250 million transacted in other countries.[3]

Other World Expansion

In India, GE Plastics-Europe formed a joint venture with Indian Petrochemicals Ltd. to compound engineering plastics and to make polycarbonate sheet and film. GE also opened a customer application center at New Delhi in 1996.

The major complex being built at Cartegena, Spain, began compounding

Cycoloy blends of acrylonitrile/butadiene/styrene (ABS) and polycarbonate (PC) in 1994, and is expanding manufacture of this rapidly growing product. To meet increasing polybutylene terephthalate (PBT) demand in Europe, GE formed a joint venture with the German BASF Corporation to make PBT resins by the continuous process that had been developed at Mt. Vernon. This plant which produces 130 million pounds per year, opened at Schwarzheide (formerly East) Germany in 1997, and its output is marketed separately by the partners. Also in Europe, GE acquired an Italian resins compounder, Resinmec, in 1997.

GEP has served the Pacific rim markets, fast-growing until 1998, with compounding capacity from a new facility in Singapore (1994), and by GEP's first China operation, which opened at Nansha in 1996. In 1999 GEP announced that it would build a compounding plant in Thailand, to come on stream in 2000.

In the Americas joint-venture compounding and marketing companies at Tampico, Mexico, and Coplen, Brazil, were converted to full ownership in 1995. And in 1997 GEP acquired an Argentinian compounder and distributor of a wide range of engineering plastics.

Another growth product line is Structured Products, which consists mainly of Lexan glazing, sheet, and film. According to GE, the continuing growth in glazing applications is driven by beauty, weight saving, and by impact strength two hundred-fifty times that of glass and thirty times that of acrylic plastic. In addition to glazing manufacture at Mt. Vernon, at BOZ, and in Italy and Japan, GEP increased capacity for these products in both Brazil and China in 1997.

In addition to many physical plant expansions around the world, GEP has reported significant increases in manufacturing productivity at both polymer plants and compounding facilities. By 1994 the initiatives to stretch capacity and reduce cycle and changeover times without major investment had freed up 500 million pounds of compounding capacity. This productivity gain, along with improved working-capital turnover and some reduction in new plant investment rate, generated large positive cash flows from 1993 through 1997 and contributed to the new highs in the ratio of operating profit to assets[1] (also refer to the financial data summary, p. 192).

Product Line Evolution

GEP's engineering plastics "turf" for the 1990s is displayed as the shaded area in Figure 12-1, on which the vertical axis is increasing *cost* and the horizontal axis increasing *crystallinity* of the polymers. A rough estimate of 1995 world market size by the broad categories Commodity ($25 billion),

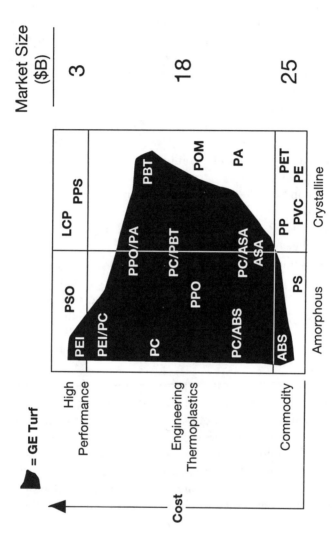

From Heuschen, "'Maximizing Technology: Strategies for Becoming a World Class Supplier,"
Presentation to SPI Meeting, Sept. 6, 1996

Figure 12-1.

Engineering Plastics ($18 billion), and High Performance ($3 billion) shows that engineering and high-performance plastics, which represent less than 5% of total tonnage, account for 40% of dollar volume. Growth rates of the premium plastics continue to be higher than those for the commodity materials.

Polymer modifications and new blends permit GEP to offer a broad range of choices for widely different applications. GE has available five polymer types, PC (Lexan), polyphenylene oxide (PPO) (Noryl), ABS (Cycolac), acrylic/styrene/acrylonitrile (ASA) (Geloy), PBT (Valox) and polyetherimide (PEI) (Ultem). Figure 12-1 also positions the major GE blend families, sometimes called "platforms," which are PC/ABS (Cycoloy), PC/ASA (Geloy), PC/PBT (Xenoy), PPO/PA (Noryl GTX), and PEI/PC (Ultem ATX).[4]

The broad markets in which these engineering plastics are being applied are shown in Figure 12- 2.[4] In these various markets GE Plastics works with both end-user customers and their molders for optimum application design, material selection, and processing technique. The program for assisting customer productivity improvement measures the savings from scrap and cycle-time reduction through this optimization.

Special product optimizations for large automotive assemblies such as instrument panels are responsible for maintaining the foothold in the automotive market. Thin wall molding materials (down to 1.5 mm), sometimes with conductivity control built into the plastic, are expanding applications in telecommunication products, such as cell phones and TV remotes, as well as in complex parts integrations for automotive assemblies. Media-storage applications like CDs, video disks, and computer mass storage are expanding as optical quality grades of polycarbonate continue to be improved. While building and construction applications have been slower to develop, there are new uses in shingles and siding. In addition, GE currently offers a high-density, mineral-filled Valox compound named Enduran as a thick plastic for countertops, sinks, and washstands.

Ultem polyetherimide plastics continue to grow, even at a premium price, and GE expects to simplify the complex PEI processes to improve the value of this technology.

To maximize the chances for success in using light-weight and safe polycarbonate glazing, initially in the rear and fixed side windows of cars, GEP and Bayer AG have announced an automotive-glazing joint venture, the Exatec Company, which has headquarters in Wixom, Michigan. The aim is to develop the materials, abrasion-resistant coatings, and process technologies to enable broad use of polycarbonate windows in advanced vehicle designs.[2]

In a different polymer arena, GE Plastics-Europe has entered a joint

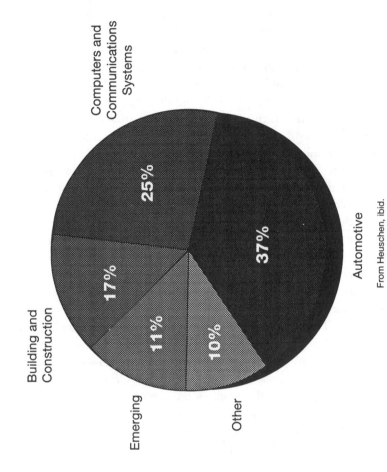

Fig. 12-2. GE Plastics, Sales by Markets 1996

research and market development venture with BP Chemicals London for aliphatic polyketones, a new class of semicrystalline thermoplastics. These polymers, developed by BP Chemicals, and by Shell Chemical, have an attractive combination of high-temperature and chemical resistance, wear, and low permeability properties. GEP brings to the development its know-how in thermoplastics and blends for applications in many markets.

Evolution of the GE Plastics Organization

The GEP organization has continued to evolve and change under Rogers. Nine profit and loss (P&L) operating components (below) now make up the total business, the first six representing the engineering resins plus related products and services as of May 1999:

The three area organizations make and sell the GE engineering resins and

	General Manager	HQ
GE Plastics-Americas	F. F. Becalli, VP	Pittsfield, MA
GE Plastics-Europe	U. S. Wascher,	
	Senior Managing Director	BOZ
GE Plastics-Pacific	M. J. Espe, President	Singapore
Polymerland	P. N. Foss, President	Huntersville, NC
Specialty Chemicals	R. Fines	Morgantown, WV
Global Sourcing &		
Petrochemicals	D. Grosman	Pittsfield, MA
GE Silicones	W. P. Driscoll, VP	Waterford, NY
GE Superabrasives	W. A. Woodburn, VP	Worthington, OH
GE Electromaterials	M. J. Abrams	Coshocton, OH

structured products in their parts of the world. Polymerland focuses on smaller customers and distributes both GE resins and a broad range of other thermoplastics. Polymerland operates on a world basis except in Europe, where the distributors report to the Europe area. Specialty Chemicals includes the phosphites and other plastics additives that were acquired with the Borg Warner acquisition. Petrochemicals represents the GEP half-interest with Fina in a styrene plant and a one-sixth interest in an ethylene facility.

Each of the areas (which GE also calls Poles) is organized according to its different size and stage of development. GE Plastics-Americas has recently changed to a seven-business product management structure in the U.S., consisting of P&L units for Lexan, Noryl, Cycolac, Cycoloy, Polycrystalline

(Valox), Ultem, and Special Products and Services, each product general manager reporting to the Americas area manager. Field marketing for all these plastics is a pooled U.S. sales and market-development organization (C. E. Crew, VP, Pittsfield). Mexico and South America have area-manager organizations.

GE-Europe area is organized functionally, and the major units are marketing and manufacturing. Country affiliates report to the area manager. The Pacific Area is organized by the major countries: Japan, Greater China, Korea, Australia, and Southeast Asia.

Product and process development are conducted by technology groups at the major plant locations. Overall technology leadership for GEP is exercised by Jean Heuschen via a dotted-line communication relationship with the technology leaders located at the plants in the United States, Europe, and Japan. The system functions more as a federation of technical centers sharing know-how, with common measurements and strong global working arrangements, rather than as a centrally controlled organization. Advanced research for engineering plastics (and silicones) is conducted at the corporate R&D Center, 60% of whose chemical research activity is now directly sponsored by GEP. The extensive application laboratory and fabrication process-development facilities in Pittsfield are available to technology leaders at all locations.

In addition to the P&L operating components, GEP staff units at the Pittsfield headquarters include (as of May 1999):

Worldwide Technology	J. M. Heuschen, VP
Global Manufacturing	A. H. Harper, VP (BOZ)
Finance	J. W. Ireland, VP
Human Resources	R. E. Muir, VP
Quality	W. Hewett
Business Development	D. Hughes
Environment, Health & Safety	M. Walsh
Legal	P. Y. Solmssen, VP
Information Management	J. M. Seral

With the world growth of GEP, personnel assignments in other countries have become more frequent. Two of the present area leaders, for example, have worked in all three areas and the other in two parts of the world organization. Multimonth as well as multiyear assignments outside a home country have become common. Higher-level management appointments will often have candidate slates from different countries.

Rogers convenes an Executive Council of about 50 GEP managers for two days each quarter to share business progress, problems, winning ideas, and management concepts. This meeting is patterned after a similar council that John F. Welch instituted at the corporation level some years ago.

6 Sigma

In GEP, as well as elsewhere in GE, projects to improve the processes for delivering products and services to customers are now addressed by multifunctional teams using a methodology called 6 Sigma. A 6 Sigma team is led by highly trained GE employees called "Master Black Belts" and "Black Belts," whose work consists of five steps: defining, measuring, analyzing. improving, and controlling selected processes. The process outcome studied might be accuracy of billing, polycarbonate batches in spec, or silicone rubber grades in-stock for customer requests. Each project's objective is reduction of process deviations outside critical-to-quality (CTQ) specifications to less than 3.4 events per million opportunities. This constitutes 6 Sigma control. Historically, GE and other well-managed companies have institutionalized 3 to 4 Sigma performance. Process control to 3 Sigma limits allows 66,807 out-of-spec events per million opportunities.[5] Taking a page from companies such as Motorola and Allied Signal, which have pioneered with the program, Welch has challenged GE managers to achieve a 6 Sigma performance across GE by 2000.

Enthusiasm for the methodology and its results is growing. In the GE 1997 annual report, Rogers noted: "A highlight for GE Plastics was our 6 sigma efforts...Our Superabrasives business...Six Sigma implementation ...yielded [additional] capacity equal to the existing plant. More than 2,200 GE Plastics employees have been trained in 6 sigma during the past two years, and we have more than 3,000 projects completed or currently in process. Total benefit for 1997 was about $137 million [of annual cost savings]."

Competition

Competitors in engineering plastics are also expanding capacity worldwide: Sumitomo Dow has completed a PC plant in Japan that is capable of producing 90 million pounds per year. Dow has announced its first polycarbonate plant in Europe and has proposed a joint venture in South Korea; Teijin Chemicals is building a large new polycarbonate plant in Singapore; and Taiwan and South Korea ABS competitors have both built very large plants. In addition, BASF is building a 330 million pounds per year ABS/ASA complex in Mexico.

GE Engineering Plastics in 1997

GE's engineering plastics, including related lines and Polymerland, recorded $5.3 billion sales in 1997, 79% of GEP total sales. Also, a *Chemical Week* article estimated that the operating income from these plastics sales was around $1,151 billion.[3] GEP is the world production leader in PCs, modified PPOs, and PEIs. It also has a major position also in PBTs, in ABS resins, and in ASA plastics.

Chemical Week estimates GE's worldwide market share for 1997 in all engineering plastics (including ABS) at around 22% by volume, followed by Du Pont at 12%; Hoechst at 11%; Bayer at 9%; BASF at 5%; Mitsubishi Gas Chemical at 4%; Dow Chemical at 3%; Asahi at 3%; Toray at 3%; Teijin at 2%; Allied Signal at 2%; and others at 24%.

*   *   *   *

Note: The following plastic applications pictures demonstrate the extraordinary utility range of GE's engineering plastics. All pictures are courtesy of GE Plastics, with permission.

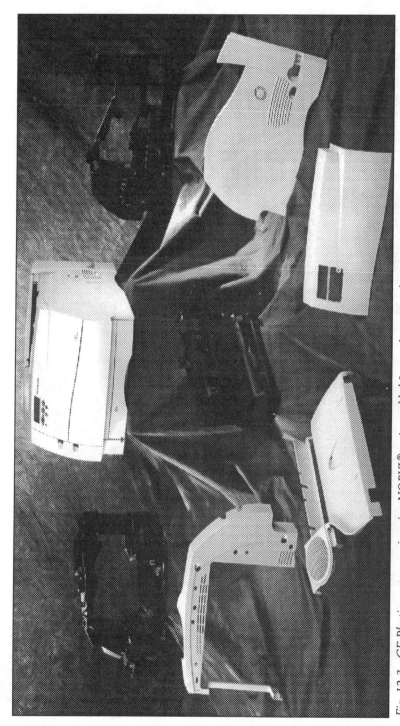

Fig. 12-3. GE Plastics custom engineering NORYL® resin enabled Lexmark to use a three-component chassis with 12 parts, as opposed to formerly 189 parts, in the new Optra S 1250 laser printer (center). GE Plastics materials used include: NORYL HM4025; NORYL SE1GFN2; and CYCOLOY® C6200.

Fig. 12-4. Rubbermaid used LEXAN® resin in its award-winning Intellivent™ food storage, microwave, and serve system.

Fig. 12-5. LEXAN® OQ1030L polycarbonate resin for the digital video disc (DVD) industry.

Fig. 12-6. GE Plastics' materials and thin-wall expertise helped Ericsson design the phone housing for the AF738 that uses CYCOLOY® C1200HF resin.

Fig. 12-7. GE Plastics worked with Textron Automotive to develop two new grades of NORYL® modified-PPO® resin for the 1998 Chrysler Concorde and Dodge Intrepid instrument panels.

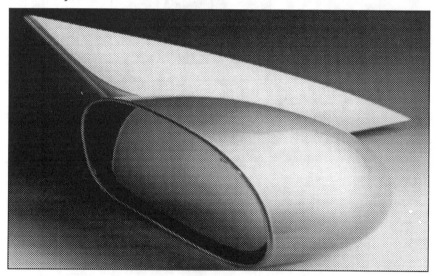

Fig. 12-8. The exterior rear view mirror housing on the 1997 Ford Taurus and Mercury Sable is the first production application of conductive thermoplastic material, NORYL GTX® nylon/PPO® resin, which eliminates the need to apply a primer coat before electrostatically painting the part.

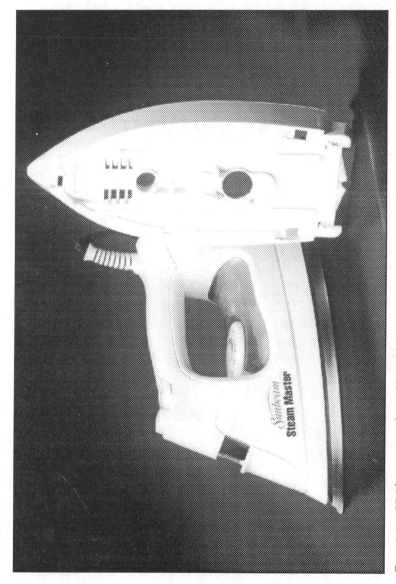

Fig. 12-9. GE Plastics new color-stable "appliance white" VALOX® CS860 resin appears in the skirts on Sunbeam's new Steam Master™ and Steam & Dry irons.

Fig. 12-10. GE Plastics LEXAN® resin is featured on the cab of the Emergency-One Concept 2000 prototype fire truck.

Fig. 12-11. Perfect Choice™, *from American sheet Extrusion corp., is a new cedar shake replica made with NORYL® resin, a flame-resistant engineering thermoplastic manufactured by GE Plastics.*

Fig. 12-12. GE Plastics' GELOY® ASA resin, part of the Weatherables™ portfolio, has allowed Alside, Inc. to greatly expand its siding color offerings and improve outdoor performance.

Fig. 12-13. GE LEXAN® bottles weigh less and are less breakable than glass.

*Fig. 12-14. GE LEXAN THERMOCLEAR® sheet has outstanding
energy-efficiency for primary and over-glazing applications.*

Fig. 12-15. Certain grades of patented LEXGARD® laminates offer high levels of containment and ballistic protection.

CHAPTER 13

People Make the Difference

Four Scientists: Eugene G. Rochow,
H. Tracy Hall and the GE diamond research
team, Daniel W. Fox, Allan S. Hay

Five Managers: Abraham L. Marshall,
Charles E. Reed, John F. Welch, Jr.,
Glen H. Hiner, Gary L. Rogers

I n the final chapter of this book we will summarize the business results as well as the management strategies that underlie GE's surprising success in the world of chemicals. But first, it is appropriate to look at GE's chemical history by focusing on its key players, some of the leading scientists and managers whose efforts and decisions made that success possible. People *do* make the difference in business, as is pointed out in this history.

KEY DISCOVERY SCIENTISTS

During the discoveries and subsequent commercial development of these complex chemical technologies several hundred chemists and chemical engineers made important contributions to the successes described. A few, though not nearly all, of these people are mentioned in this history. Of these, four individuals, one with a team, stand out for having made the discoveries that opened four technology areas of great importance. From each of these technologies GE established a successful chemical business or product line.

EUGENE G. ROCHOW (1909–)[1]

Eugene G. Rochow's discovery of the copper-catalyzed reaction of methyl chloride with silicon to form methylchlorosilanes, and particularly dimethyldichlorosilane, was the early defining event for GE in silicones. He had previously prepared methyl and phenyl silicone resins using organochlorosilane intermediates prepared from silicon tetrachloride and a Grignard reagent, a process that neither Abraham Marshall nor Winton Patnode believed would lead to economical silicone polymers. Rochow persisted in seeking a direct process and made his discovery in May 1940.

This breakthrough stimulated Marshall to expand the Research Lab silicone project greatly, which led to other silicone polymer discoveries, notably fluids and elastomers. Corning and Dow Corning chemists were developing similar silicone polymer forms, but Rochow's direct process was a unique discovery. Absent this invention, it is clear that GE would not have entered the silicone business; and the silicone project discoveries were a significant factor in GE's 1945 decision to form a chemical division.

Rochow's book, *An Introduction to the Chemistry of the Silicones* (Wiley, 1946), developed from notes with which he indoctrinated new GE chemists to the silicone project, was the first modern publication of the new field.

In 1945 GE asked Rochow to transfer to nuclear power research, which GE was entering at the Hanford, Washington, atomic site, and subsequently at the new Knolls Atomic Power Laboratory near Schenectady, New York. When, in 1948, the research objective of these laboratories shifted from commercial atomic power to submarine propulsion, Rochow, a Quaker and a pacifist, left GE to join Harvard University, where he was later recognized for his effective teaching of inorganic chemistry. He received many honors during his career, including the Perkin Medal of the Society of Chemical Industry. This honor, in particular, brought international recognition, including a guest professorship at Innsbruck University in Austria, and honors from other European universities. He became Professor Emeritus from Harvard in 1970, and has remained active in the field by attending international organometallic technical conferences.

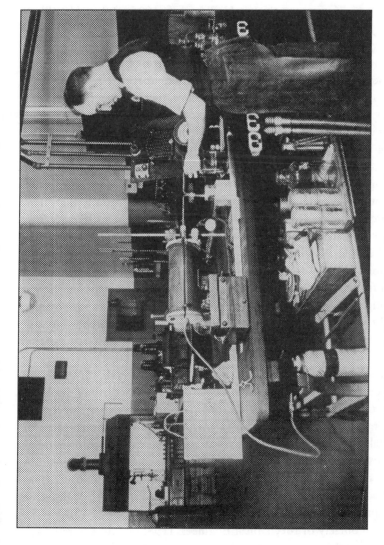

Fig. 13-1. Eugene G. Rochow, silicone pioneer researcher and inventor of the direct process for methychlorosilanes. ca. 1940

H. TRACY HALL (1919–) and the GE diamond research team

H. Tracy Hall designed the high-pressure carbide "belt" apparatus and in December 1954, chose the cell ingredients that first yielded synthetic diamond. His successful experiments were the culmination of a team effort that began in 1951 and that had seen important contributions also by Francis P. Bundy, Herbert M. Strong, Robert H. Wentorf (1926–1997), Harold P. Bovenkerk, James E. Cheney, and team leader Anthony J. Nerad. The diamond research group was part of Marshall's chemistry jurisdiction, but the targeted program had been launched at the instigation of C. Guy Suits, then Vice-President and Director of the Research Laboratory.

GE promptly formed a commercial section within the Metallurgical Products Department to exploit the research discovery by process, product, and market development. Bovenkerk and Cheney joined this Detroit-based effort. Hall left GE and joined Brigham Young University as Director of Research, continued research in the high-pressure field, founded the Megadiamond Company, and is Chairman of Novatek, a "Super Materials" company. Bundy, Strong, and Wentorf continued high-pressure, high-temperature research at the Research Lab for the balance of their GE careers.

The GE scientists received many honors for their accomplishments. The American Chemical Society (1972) gave Hall a special award for creative invention for being the first to discover a reproducible reaction system for making synthetic diamonds from graphite, and for the concept and design of a super-high-pressure apparatus that not only made the synthesis possible, but brought about a whole new era of high-pressure research. He also received the Chemical Pioneer Award of the American Institute of Chemists (1970).

The GE diamond development team of Bundy, Hall, Strong, and Wentorf received the Research Medal of the American Society of Tool and Manufacturing Engineers (1962), the Engineering Materials Achievement Award of American Society of Metals (1973), and an international prize for new materials from the American Physical Society (1977). At this society's 100th anniversary meeting (1999), *Nature* magazine presented "A Celebration of Physics" special issue. In it, the editors' selection of 22 from its most important papers of the past century included the 1955 Man-Made diamonds paper of the four GE scientists (see reference 21 of Chapter 5). The American Society of Mechanical Engineers (1988) commemorated the scientist team and the Research Lab with a plaque and exhibit at the Schenectady Museum.

Bundy received the Roozelboom Gold Medal of the Netherlands Academy of Arts and Sciences for extending the carbon-phase diagram

Fig. 13-2 H. Tracy Hall, first to discover a reproducible reaction system for making diamonds from graphite. ca. 1955

(1969, Amsterdam), and the Bridgman Gold Medal of the International Association for Advancement of High Pressure Science and Technology (1987, Kiev). Wentorf's honors included the ACS Ipatieff Prize (1965) and an Industrial Research Institute Achievement Award (1970). In addition, he became a member of the National Academy of Engineering in 1979.

DANIEL W. FOX (1923–1989)[2]

After joining GE in 1953, Daniel W. Fox made his polycarbonate (PC) discovery as a serendipitous byproduct of his first work assignment. He was put on a Research Lab chemist team that was developing a new wire enamel for high-temperature service. Seeking ways to improve its hydrolytic stability, he tried several syntheses to form a polymer containing carbonate units, an idea stemming from one of his university postdoctorate experiments. His transesterification reaction of bisphenol A (BPA) and diphenyl carbonate at high temperature and under vacuum, distilled off phenol and yielded a glob of high melting plastic. Its almost unbreakable character interested the Research Lab chemists, though it made no contribution to the wire enamel project that was their priority.

Fox was coinventor of the successful wire enamel product, trade-named Alkanex. But he returned to his Schenectady Works Lab consulting position where he continued further polycarbonate experiments on bootleg time. After Alphonse Pechukas initiated a PC development program in the Chemical Development Operation, Fox joined the team in 1956. His "melt process," could not be scaled up at that time to make a usable plastic because equipment was not then available to deal with the high-temperature, high-viscosity mixing requirements under vacuum. Undaunted, the Pittsfield, Massachusetts, development team worked out a totally different synthesis, using phosgene and BPA, to create the desired polymer for a molding material. This was trade-named Lexan, and was first offered to U.S. customers in 1957. This GE work, it was by then learned, paralleled similar discoveries by Hermann Schnell in a Bayer German laboratory.

Fox helped develop the Lexan process improvements that were later incorporated at the Mt. Vernon, Indiana plant. He then rejoined the Chemical Development Operation where he became the important first champion for another GE thermoplastic, polyphenylene oxide (PPO) and its ultimate use in the Noryl blend. Early in this job he persuaded a chemical engineering candidate from University of Illinois, John F. Welch, Jr., to join GE and the PPO project. Following successful commercial launch of the PPO/Noryl venture and Welch's appointment to manage the renamed Plastics Department, Fox and Welch teamed up to add Valox polyester thermoplastic (PBT) to the GE product line.[2]

Dan Fox thus discovered Lexan PC plastic for GE, and he personally made key contributions to both Noryl and Valox. During his 35-year career with GE, he made important contributions to the company's success with engineering plastics through his work and leadership. With William F. Christopher, he published an early book describing modern commercial PCs

Fig. 13-3. Daniel W. Fox, with Plastics Hall of Fame award, 1976; GE's Lexan discoverer, and also champion for Noryl modified-PPO; discoverer of and proponent for a chlorine-free polycarbonate process.

(Reinhold, 1962). While GE and competitors built PC interfacial polymerization plants around the world, he kept alive the vision that a PC process without solvent or phosgene could be developed. Before his retirement in 1988 he was pleased to visit Japan and the GE-Mitsui world-first plant to make PC by the melt process he had discovered in 1953.[3]

Fox's associates remember him as a great experimenter and important mentor: Popkin Shenian, a Pittsfield associate for 25 years: "Dan has made great contributions to this business, but one of the greatest was hiring a cadre of outstanding scientists. We call ourselves graduates of Fox U., and we are all very proud of it."[4]

Edward Peters, Selkirk, NY: "In spite of all the modern equipment, we were unable to make it work. After two months of waiting for us, Dan Fox became impatient. He walked into our laboratory, took a few pieces of crude equipment and a beaker and made the polymer on his first attempt...He's either a great experimentalist, or very lucky."[4]

Fox received many honors for his career accomplishments, including a place in the Plastics Hall of Fame and the International Award of the Society of Plastics Engineers (SPE). He was also a member of the National Academy of Engineering. The city of Pittsfield honored his community contribution by naming their airport approach Dan Fox Drive.

165

ALLAN S. HAY (1929–)[5]

When Allan S. Hay arrived at GE in 1955 he already had an unusually broad laboratory experience from his chemistry studies at Universities of Alberta and Illinois, plus summer work at Canada's National Research Council and with Du Pont. At the GE Research Laboratory his interest in the oxidation of organic molecules led first to a high-yield process for making terephthalic acid from p-xylene, a finding GE did not follow up. Hay then experimented with oxidation of phenols, and finally of 2,6-xylenol. In 1956 his catalyzed oxidation of this monomer yielded an unusual polyphenylene oxide polymer. Labeled PPO by GE, it had a higher melting point and greater hydrolytic stability than Lexan polycarbonate, which GE was then putting into the market.

Hay's early papers and later issued patents attracted world attention and led to a GE joint evaluation of PPO with the Dutch firm, AKU, as a fiber and as a molding compound. Hay and his associates also demonstrated a monomer synthesis and then oxidative polymerization of 2,6-diphenylphenol, to a polymer they called P3O. Neither PPO nor P3O fibers proved useful, but GE's molding compound interest continued.

GE committed to a full-scale PPO plant in 1964. But a fatal characteristic of the polymer was severe degradation at its high melting temperature during injection molding. This weakness was never overcome, but a versatile, good-performance molding plastic was created by mixing PPO with several parts of low-cost, high-impact polystyrene plus some rubber. Hay's unique PPO discovery plus the Noryl mixtures technology, marketed along with Lexan and later Valox, gave GE Plastics a world leadership product in the engineering thermoplastics.

Like Fox, Al Hay was a great experimentalist. A Research Lab associate recalls Hay's morning laboratory routine:[4]

...about 6:30–7:00 in the morning...Alan would come in and pick up JOC, or Chem Abstracts, or JACS and read. And...he would actually just clip out the (Chem Abstracts) references that interested him and paste them on file cards...If it was something he wanted to spin off real fast, he would have six reactions going by 9:00 A.M...And he had a technician that came in at 8:00 and he'd have enough work for that technician to do all day long.

Hay managed the Research Lab group that developed a polyimide synthesis first proposed by Joseph G. Wirth, through many steps to a moldable polyetherimide thermoplastic. This was a 10-year effort and the expensive (ca. $5/lb) Ultem final product, now made in the Mt. Vernon plant, has exceptional properties. Ultem applications are growing and plant capacity is being added.

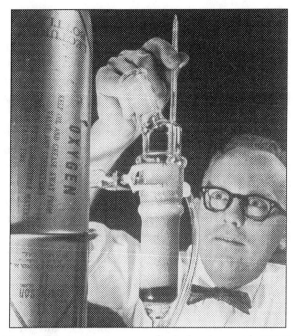

Fig. 13-4. Allan S. Hay, inventor of polyphenylene oxide (PPO) polymer, the precurser to Noryl thermoplastic; and leader of research yielding Ultem polyetherimide thermoplastic. ca. 1964

With respect to Wirth's early work on imides, Hay described a GE Research Lab philosophy for open inquiry that had earlier attracted him to join the company:[5]

Essentially we are always happy to maintain an exploratory effort on almost anything as long as novel ideas are being pursued...If you are doing something that is on the leading edge, great; maintain it. Here was a case where there was some literature that indicated that you could displace a nitro group. Thus if you could design a bifunctional molecule you had the potential for making a polymer..."

In later years Hay managed a GE Research Lab chemistry effort of over 200 technical people. Then in 1987 he joined McGill University as holder of the GE/NSERC Polymer Chemistry Chair. He holds 87 U.S. patents, has authored more than 180 publications, and has received many professional honors. Hay is a Fellow of The Royal Society of London. He received the International Award of the Society of Plastics Engineers (SPE), the Achievement Award of the American Institute of Chemists (AIC), and in 1985, the Carothers award. Hay currently (1999) holds the Tomlinson Chair in Chemistry at McGill.

KEY MANAGERS

Outstanding scientists like those we have described explore a field and make discoveries, while operating management decides which opportunities to follow and how to pursue them. In following up new polymer chemistry the financial commitment risk for product, process, and market development, plus later plant construction are far greater than the research costs. And the important early management decisions must be made in great uncertainty, without clear knowledge even of technology futures, and certainly not of market size, competition, and eventual costs and pricing.

Companies don't make decisions, but people within them do. GE's successful strategies are the initiatives of many leaders. From hundreds of significant management leaders in GE's chemical history, five stand out for the importance, breadth, and duration of their contributions.

ABRAHAM L. MARSHALL (1896–1974)

Abraham L. Marshall, who joined the GE Research Laboratory in 1926, was put in charge of its Insulations Section in 1933 and led the Lab's chemistry and insulations research for 28 years, during great growth and with many different titles, until his retirement in 1961. Marshall's own scientific career focused on physical chemistry phenomena, but in the management role his interest shifted to polymer research, particularly those chemistries showing promise for superior electrical insulation. In the 1930s GE's success with extruded vinyl polymer insulation, Flamenol, and with the breakthrough Formex wire enamel established the chemical section's reputation in the company.

In concert with Patnode, Marshall's vision extended to GE participation in chemical manufacturing, if research could provide the opening. Silicone polymers and Rochow's direct process were discoveries that opened a large-scale opportunity, and Marshall seized the initiative to build staff and conduct a major research program. With Research Lab Directors William D. Coolidge and Suits, he helped persuade GE top executives to pursue silicones as a business.

Herman A. Liebhafsky describes Marshall's personal characteristics:[6]

"He was energetic, strong-willed, reticent, complex, and gifted with intuition and insight. He had a memory so good that he seldom needed to turn to books and could operate largely on the basis of what he had been told: he was a superb listener, with a quick mind that never rested...He normally expected others to meet the high standards he set for himself. In the pursuit

Fig. 13-5. Abraham L. Marshall and John R. Elliott examine an enamelled wire sample. ca. 1957

of chosen objectives, scientific or business, he could be unsparing of himself and others; he always kept first things first."

Marshall attracted a great staff to the Research Lab early on, including Patnode, Rochow, and Charles E. Reed, each a major contributor to GE silicone research and development. Marshall managed several modes of Research Lab effort easily. Some assignments went to pathfinders, who were encouraged to explore, frequently on their own initiative. Other projects were assigned to teams. And Marshall's organization was always ready for chemical troubleshooting assignments anywhere in the company. He had complete faith in individual creativity: Reed once asked him how many people would be needed to assure a particular polymer discovery. Marshall's reply: "Just one, the right one."

Research Lab chemistry output on Marshall's watch following silicones was also impressive: Permafil electrical varnishes, Alkanex wire enamel, Vulkene irradiated polyethylene wire insulation, synthetic diamond and Borazon superabrasives, Lexan polycarbonate thermoplastic, polyphenylene oxide and Noryl thermoplastics, and a 2-6 xylenol synthesis.

Dan Fox later spoke of Marshall:[2]

"Probably the greatest person and inspiration I worked with was...Abraham Lincoln Marshall. Abe Marshall was a madman from the point of view that he ran everywhere in a hurry. He was totally safety unconscious. He made commitments to people when he shouldn't. He did all sorts of things but I liked him because he really had the courage of his convictions. He went out and sold this product, this Class B wire enamel, before we had it. He did a great job of selling it. He told everybody, "We will have it. We will deliver."

CHARLES E. REED (1913–   )

Charles E. Reed's 1991 National Medal of Technology award from President George H. W. Bush honored his contributions during a 37-year career with General Electric. Reed is widely recognized for his leadership of GE's chemical businesses during their early years and the growth to success of silicones, diamonds, and engineering plastics.[7]

Reed joined the Research Laboratory in 1942, where he addressed distillation of methylchlorosilanes and the direct methylchlorosilane process. He invented the application of fluidized-bed technology to Rochow's direct process reaction. When GE's Chemical Division was formed in 1945 he became responsible for process design of the Waterford, New York silicone plant and he managed process development groups in Pittsfield and Waterford. The addition of a Pittsfield new-product development laboratory in 1950 increased his responsibility.

Reed became the first general manager of an independent Silicone Products Department in 1952, overseeing its growth to breakeven and then to respectable profitability, while expanding both R&D and technical service. He kept his own office in a farmhouse until a suitable silicone R&D laboratory was built. In early 1959 he became general manager of the larger Metallurgical Products Department, which included the diamond product section.

GE promoted Reed to Vice-President in charge of the Chemical and Metallurgical Division in 1960. During the next years Lexan grew to a profitable business, and he championed Noryl through its successful launch. Silicones continued rapid growth and achieved U.S. market leadership in sealants. In 1967 he moved up to Vice-President and Group Executive with continuing responsibility for the chemical businesses, but also for electronic components, small motors and controls, and medical X-ray equipment. By 1972 Valox had been successfully introduced, GE was marketing engineer-

Fig. 13-6. Charles E. Reed, Senior Vice-President, Technology, ca. 1977

ing plastics in Europe, was manufacturing at Bergen op Zoom, and Japanese distribution had been established. The diamond line had become a world success.

Above all, Charles Reed learned from the GE silicone experience and used that understanding not only to shape his own management decisions, but also to teach his associates and superiors. His technical understanding and enthusiasm were key to excellent relations with R&D Center scientists, who had considerable latitude in where to direct their own research. It was a company article of faith that Reed and his associates attracted more than their share of Research Lab effort in support of their ventures. Organizations under Reed leadership had open flow of information. When problems arose, he was never interested in who was responsible, but only in what needed to be done going forward. He welcomed open discussion and even direct criticism of his policies without malice to any individual. He was a long-range thinker on all subjects.

Reed was promoted to Senior Vice-President on the corporate staff in 1971. His counsel concerning the chemical and nuclear businesses, technology in general, and in corporate strategic planning was important during the leadership of CEOs Fred J. Borch and Reginald H. Jones. He retired from the position of Senior Vice-President—Technology, in 1979. In a farewell address to the 1979 GE Management Conference, Reed acknowledged the top GE corporate support for the chemical businesses:

"Nothing—repeat nothing—saved these foundling businesses except a favorable climate at the top of the company..." But he added, "Another thing I've learned...much depends on whether you want money for capital investment or money for current expenses...Capital is a *soft* touch; Expense is *hard* (on earnings!)"[8]

Reed was considerate of and helpful to all of his associates with respect to their careers. And he is proud that the company's present leadership includes both Welch, GE's Chief Executive Officer, and John D. Opie, a Vice-Chairman, each of whose GE service began in the Chemical and Metallurgical Division.

John F. Welch, Jr. (1935–   )

John F. Welch's direct association with and management of GE's chemical businesses spanned 17 years, from 1960 to 1977. As a chemical engineer with the polyphenylene oxide project, he built the first pilot plant, designed a purification process, and later the synthesis demonstration for the 2,6-xylenol monomer. His leadership ability gained him the job as manager of the Polymer Products Operation, which built the Selkirk plant and took the PPO/Noryl product to market. Welch and his team changed the PPO plan to the very different Noryl mixtures of PPO with polystyrene, a shift that saved the program from failure and created opportunity to develop a new segment of the engineering plastics market. Noryl was successfully launched worldwide under his direction.[9]

When the Polymer Products Operation was combined with Lexan polycarbonate and the phenolics product lines in a renamed Plastics Department, Welch became its general manager in 1968 at age 33. His special impacts on the business were threefold: (1) increased marketing and market development effort; (2) worldwide market penetration, employing commercial development centers and overseas manufacturing; and (3) speedy competitive action, as exemplified by the Valox plastic development and marketing. By the mid-1970s GE's U.S. market-development success had been transplanted to Europe and Japan, and each of GE's engineering

172

Fig. 13-7. John F. Welch, Jr., Chairman and Chief Executive Officer, General Electric Company, 1981 to the present.

plastics—Lexan, Noryl, and Valox—was a world market leader.

On the one hand, operations under Welch's leadership were fast-moving, aggressively competitive, and resistant to corporate directives. On the other hand, their appropriation requests were well-justified and their strategic-plan documents well thought through.[10]

Welch was promoted to Division Vice-President for all chemical products in 1972. By 1974 he had advanced to Group Executive, still with responsibility for chemicals, but also for three additional divisions. He became Sector Executive for Consumer Products and Services (1977), Vice-Chairman (1979), and has been Chairman and Chief Executive Officer from April 1981 to the present day.

During Welch's Chief Executive years the company has continued to support rapid growth of the chemical businesses. Annual reports show that from 1985 through 1997 GE Plastics plant and equipment (P&E) expenditures (not including the $2.3 billion ABS acquisition) amounted to $7.5 billion, a surprising 26% of the P&E investments for the entire GE company.

In a recent interview, Welch points to two important perspectives from his plastics business experience that influenced his rise to the top and some of his later chief executive initiatives. The first: How critically important is the *quality of players* on a team. He notes that particularly in a small organization a strong contribution from every person is needed; and if lacking, the manager needs to make changes earlier rather than later. This focus on personnel performance, development and rewards is now expressed throughout GE during "Session C" personnel reviews, covering the entire corporation.

The second perspective is *globalization*: the opportunity, and necessity, of competing effectively for business worldwide. Welch led GE Plastics from a U.S. orientation to a successful worldwide business, thereby multiplying the returns from the technical and marketing achievements in the home country. Overseas management assignments were not sought in GE at one time, but Welch insisted on recruiting the best people for foreign assignments and also providing advancement channels for foreign nationals in the organization. He says that GE Plastics and GE Medical Systems are today the most globally oriented of GE's businesses.[11]

GLEN H. HINER (1934–   )

Glen H. Hiner, a 1957 electrical engineering graduate of the University of West Virginia, worked in GE's electrical businesses. He managed a controls operation in San Leandro, California, a vertical motor business in San Jose, California, and then became general manager of a large fractional-horsepower motor department in Ft. Wayne, Indiana. Hiner joined GE Plastics in 1977 as head of the European operations, returning to Pittsfield in early 1979 to become Vice-President and General Manager of the Plastics Division. He managed engineering plastics as it grew to Group status, and he became Senior Vice-President in 1983. GE's other chemical businesses were added to his responsibility in 1986. He conducted the acquisition of Borg Warner's (BW) ABS business in 1988. During his 13-year leadership, GE's engineering plastics sales grew from around $700 million to $4 billion sales (including $1.6 billion sales from the BW acquisition).

Major European manufacturing facilities added during his leadership included PPO polymer at Bergen op Zoom and the start of a new manufacturing site at Cartagena, Spain. In the U.S., an Ultem plastic line was installed at Mt. Vernon and a major new Lexan and BPA facility constructed

Fig. 13-8. Glen H. Hiner, leader of GE's engineering plastics for 13 years, and of all GE Plastics, 1986-1991.

at Burkville, Alabama. A polystyrene plant venture with the Huntsman Company was built at Selkirk.

In Japan, joint ventures were initiated with Mitsui Toatsu and Mitsui Petrochemicals for 2,6-xylenol and PPO polymer. Success with these projects led to Japanese BPA and PC plants, also with Mitsui Petrochemicals.

At the Pittsfield headquarters location three major facilities were added: a new laboratory plus headquarters office; a large Polymer Processing Development Center, and the Living Environments plastics concept house.[9, 12]

Hiner left GE at the end of 1991 to become the CEO of Owens Corning Corporation.

175

GARY L. ROGERS (1944–   )

Gary L. Rogers joined GE's Financial Management Program in 1966 from Florida State University. Before becoming President and CEO of GE Plastics in January 1992, he held several senior positions, most recently as Chief Executive of GE Appliances.

Although GE Plastics was experiencing a significant earnings decline during the world recession, expansion of PC capacity continued at all locations: at Mt. Vernon and Burkville in the U.S., Bergen op Zoom, the Netherlands, and Chiba, Japan. A new 285-million-pound PC unit at Cartagena, Spain, that came on-stream in 1999, is the first GE polymer production at that location. Since 1992, GE Plastics has doubled its world PC capacity. The company recently announced (1999) it will further expand the plant in Spain with a second polymer line by 2002. Both the Japanese and Spanish PC plants employ unique new processes that eliminate both methylene chloride solvent and phosgene, and that are expected to deliver lower production costs. In other polymers, Valox (PBT) capacity in Europe has been expanded by a joint venture plant with BASF in Germany. Cycolac and Cycoloy capacity is being increased in the U.S., and Geloy ASA weatherable plastic is being added to the ABS plant at Washington, West Virginia. GE Plastics has continued expanding compounding capacity around the world: Cycolac and Cycoloy in Spain; all products in Brazil and Mexico; a compounding and Lexan glazing joint venture in India, and GE Plastics' first manufacturing unit at Nansha, China.

Polymerland, the plastics distributor organization, has been expanded in the U.S., across Europe, and in India. Rogers has redefined the Polymerland role with relation to direct GE sales and to distribution of many plastics not made by GE. Sales via this channel are estimated at $1 billion worldwide.

Two potentially important associations with Bayer AG of Germany have been initiated: an automotive glazing joint venture to develop polycarbonate sheet and coatings technology for automotive side and rear windows, and a silicones venture that will combine the manufacturing, technical, and commercial resources of GE and Bayer Silicones in Europe, Africa, and the Middle East.[13, 14]

Recent GE annual reports cite many productivity improvements and plant-capacity increases through improved cycles, reduced changeover time, shared global best practices, and a product-simplification process that reduced the number of engineering plastics grades. The success of these initiatives yielded capacity increases with minimum investment, and the bottom line shows impressive improvement in the ratio of both profit to sales and profit to assets. (See Fig. 9-1, page 119, and Financial Data on page 192.) The recent ratios compare very favorably with leading U.S. and world chemical companies.[16]

Fig. 13-9. Gary L. Rogers, President and Chief Executive Officer, GE Plastics, 1992 to the present.

In the 1997 annual report Rogers sums up his GE Plastics management outlook: "I think we've been unique in our global position, we have the ability to place people and sales facilities throughout the world...We would like to be thought of as unique in the eyes of customers. This is what's important to us, this is the way we ought to measure how unique we are."[15]

CHAPTER 14

Summation

How Big an Achievement?
How Attained?
Nine Strategies

General Electric's 1998 annual report cites GE Plastics sales of $6.6 billion and operating profit of $1.6 billion.[1] Engineering plastics (ca. 79%) and silicones (ca. 15%) make up most of the sales, the balance being superabrasives and industrial laminates. A profit contribution estimate for 1997 put engineering plastics and related lines at 78%, silicones at 13%, and superabrasives at 9% of the total.[2] World leadership in polycarbonate plastics (Lexan) and in modified-polyphenylene-oxide plastics (Noryl) are the largest factors in GE's preeminent engineering plastics position.

In this history we traced the growth of GE's chemical businesses from sales of $50 million in 1948 (about half internal) to $6.6 billion in 1998 (0.3% internal), while the earnings contribution grew from negligible in 1948 to $1.5 billion operating profit. Within GE the chemical product lines contributed 12% of company earnings in 1998, exceeding that from either turbines, appliances, lighting products, or NBC broadcasting.[1]

We have seen that GE's chemical business success was not the result of a top-down corporate program. Company officers above the level of chemical top managers rarely interfered with business initiatives and strategy bubbling up from within the chemical organization, though they continually applied pressure for good annual results. Key to this independence, of course, was the credibility of GE's chemical leaders as sales grew, competence developed, and profitability improved.

In the following I cite nine *initiatives* or *strategies* that GE chemical managers chose, which contributed to the remarkable results. While these strategies were never explicitly decreed, they were discussed and debated, and I believe these summarize GE's successful chemical management precepts.

Invest Heavily, but Only When Research Discoveries Provide a Leadership Opportunity

With one acquisition exception, GE's chemical business growth was based on investment following internal discovery and development. Some related technology was purchased or licensed and several joint ventures were entered where technology or culture would create a more viable enterprise. But in the main, it has been a grow-from-within innovation strategy. Patient support of research and development yielded pioneer (though not always unique) discoveries by GE chemists in five areas. In time order, these were:

Silicone polymers plus the direct process for making methylchlorosilanes, 1938–1944;

A process for making industrial diamonds, 1951–1954;

An aromatic polycarbonate (PC) thermoplastic with some outstanding properties (Lexan), 1953–1956;

A polyphenylene oxide (PPO) polymer from a xylenol monomer and blends therefrom for useful thermoplastics (Noryl), 1956–1960;

A complex polyetherimide thermoplastic (Ultem), 1969—1979.

Arguably, the discovery leading to the greatest commercial success was the aromatic polycarbonate (Lexan), followed by the modified PPO/polystyrene product family (Noryl). The GE PC discovery was made independently of, but coincident with, similar polymer observations in Bayer laboratories, the latter prevailing in the first patents issued. The PPO polymer discovery proved unique in the world. Together, Lexan and Noryl products gave GE pioneer market opportunities in the then new field of premium-performance, premium-price engineering thermoplastics. GE added polybuytlene terephthalate (PBT) polyesters (Valox) and successfully marketed thermoplastics from these three polymer technologies to establish a large leadership niche in the world chemical industry.

GE Research Laboratory chemists provided early polymer discoveries of silicone resins, fluids, and rubber, plus a superior process for making methylchlorosilane intermediates. While the direct methylchlorosilane process was a unique contribution, competitive silicone product and market development by the experienced management team at Dow Corning, with a time-equivalent start, totally outdistanced GE's commercial silicone effort. GE entered the market-development race and took the big risk of building a silicone plant that opened in 1947, but the business result was very disappointing compared to that achieved by the competitor. In time the GE Silicones organization gained competitive competence, leadership in some areas, and satisfactory financial results, but the opportunity had passed for early success and market leadership.

The silicone opportunity was an important factor, however, in GE's 1944 decision to combine this project and some disparate chemical and semifinished materials businesses into a Chemical Division. Most importantly, the silicone experience provided critical management lessons for how similar opportunities should be managed. These lessons were well-learned and the strategies effectively applied, first with the synthetic diamond discovery and then with engineering plastics. GE "went to school" on the silicone experience.

The discovery of a high-pressure, high-temperature process for making synthetic diamond, announced to the world in February 1955, was the result of a 4-year targeted team effort by scientists of the GE Research Laboratory. This scientific quest had been pursued over two centuries by dozens of scientists around the world. The unique technical achievement gave GE a new opportunity for commercial venture with a novel process and a new material. Chemical and Metallurgical Division management, to which the diamond discovery was assigned, took full advantage of the opportunity.

Commit Product Development, Market Development and First Plant Capacity Rapidly

Of management omissions by GE in silicones none were more critical than delay in assigning *focused general management* and in fielding an *adequate sales force*, the latter to contact potential customers, follow up samples, close orders, and feed back application and competitive information. Effective product R&D should respond to the requirements of paying customers and must also be sensitive to the performance of competitors. Speed is important from the outset, because market share with a new technology is established early and becomes difficult to change in later years.

With the diamond opportunity GE promptly assigned commercial responsibility to a dedicated product section with its own manager, housed within a larger department that had related technologies. Research Laboratory work continued, but the new organization rapidly set out to find whether the high-pressure, high-temperature technology would yield a synthetic product that could compete with natural diamond boart. In three years time, process development and customer feedback showed that the synthetic could surpass natural-diamond in grinding wheels. So a new and profitable business was successfully launched, and at the same time the next great diamond product target (for saws) was identified.

With both Lexan and Noryl thermoplastics a dedicated commercial organization, the Pittsfield-based Chemical Development Operation, followed up research discovery with the early phases of product, process, and market development, including suitable pilot-plant and semiworks capacity for customer evaluation material. GE set up and staffed application development facilities to be able to test plastics as customers would and to demonstrate how these new materials could be molded satisfactorily. To reach the U.S. market first, GE made the initial Mt. Vernon, Indiana, Lexan plant investment with an available, but predictably costly process. The Selkirk, New York, plant for Noryl was also rushed, even before the PPO polymer had proved its utility.

With these two plastics families and the added Valox polyester line for solvent resistance, GE's chemical leadership correctly perceived the enormous long-term potential for premium-performance thermoplastics, as well as the marketing process needed to accelerate the applications. Market-development specialists were added in the field and at headquarters to more rapidly develop new applications with customers. In 1970 the first of an eventual 12 Commercial Development Centers was placed near U.S. automotive users of engineering plastic parts. Market development personnel, and these Centers, were a big factor in the success of GE engineering plastics around the world.

With the new product opportunities, diamonds excepted, GE made the first large plant commitments facing major uncertainty and high risk. New plants at Waterford, New York, Mt. Vernon, Indiana, Selkirk, New York, and Bergen op Zoom, The Netherlands, were each a leap of faith when first proposed, but all have been successful and each has expanded enormously beyond the first design.

Pursue World Position Early

In silicones GE took no early initiative toward manufacturing outside the U.S. and Canada. International General Electric (IGE) had independently made commitments for patent licensing with three major chemical companies in Europe, and licenses requested later under Japanese patents were negotiated with two companies in that country. Silicone export sales were handled by a pooled sales organization under IGE, but specialist sales reps were not located outside the U.S. GE's first offshore joint venture in silicone manufacturing began in 1971 when GE Silicones purchased a 49% interest in Toshiba Silicones. By that time, GE had fallen far behind Dow Corning in world participation.

The GE diamond management team, however, took a world view from the beginning, stimulated by phone calls and personal visits from diamond-customer executives in several countries. Specialist selling overseas directly responsible to U.S. management was put in place, as was clear authority for diamond patent and licensing decisions. Success eventually yielded greater sales outside the U.S. than within. Aggressive pursuit and enforcement of patents worldwide resulted in a high hurdle for competitors and substantial licensing income.

With engineering plastics a special opportunity for world participation came with the unique polyphenylene oxide technology and Noryl products. The AKU (Dutch) interest in a European development joint venture, which had not yielded useful fiber technology from PPO or P3O polymers, was converted to full GE ownership in 1969, land in The Netherlands was purchased at Bergen op Zoom (BOZ), and European marketing plus Noryl compounding began. Lexan sales in Europe and compounding at BOZ followed when additional Bayer licensing was negotiated. Polycarbonate was the first polymer manufacture at BOZ, followed later by PPO. The GE Plastics organization marketed these thermoplastics aggressively, with support from Commercial Development Centers, first in Europe and then soon around the world.

In Japan, a 1971 joint venture for marketing and compounding all GE's engineering plastics proved effective. Subsequent joint ventures to make PPO polymer and its xylenol monomer were put in place with a different partner. Lexan polymer and bisphenol A (BPA) are now also manufactured in a joint Japanese venture. And currently the second GE European polycarbonate plant is opening at Cartegena, Spain, already home to Cycolac (ABS) and Cycoloy (ABS/PC) compounding.

In addition to the *polymer* manufacturing locations outside the United States, GE established *plastics compounding* facilities in Canada, Brazil, Mexico, Australia, India, Singapore, South Korea, and China.

Expand Capacity Ahead of Demand with Improving Process Technology

Fundamental to long-term success in rapid-growth markets is keeping plant capacity ahead of demand. Because integrated chemical-plant capacity requires 18 to 30 months to put in place, management commitments must be made well ahead of capacity runout. In engineering plastics GE successfully avoided market-share loss that would have resulted from insufficient plant capacity. Capacity-stretch programs, added process lines at existing plants, and sometimes new green-field plants, such as the more recent polycarbonate units at Burkville, Alabama, Chiba, Japan, and Cartegena, all played a part. Major investment approvals by the company's top executives and the board of directors became easier as the organization continued profitable growth and could point to satisfactory results with previous plant expansions.

Continuing process development and improvement was also achieved in silicones, diamonds, and engineering plastics so that each iteration of capacity expansion was more productive than its predecessors.

Seek Product Line Expansion and Selective Forward Integration

GE did not insist that customers adapt to available product technology. Industrial customers for premium materials prefer optimized products, ready for their particular use. A successful material supplier must offer continuous product improvements, and a balance needs to be struck between the manufacturing economy of fewer grades and the competitive desirability of a tailored offering for each application. I believe the record indicates GE chemical managers sought product improvements and offered market-driven grades expansion as readily, or more so, than the competition. For example, the Noryl family of molding compounds was the industry pioneer example (1965) of blending dissimilar polymers to offer a unique properties combination.

Significant forward integration successes include Lexan glazing, sheet, and film, as well as the broad line of GE silicone sealants. In each case an independent customer manufacture level existed, or could easily have been developed, to perform the final process, packaging, and marketing function; but GE managers successfully coped with the added manufacturing and marketing complexity, thereby increasing value added and becoming both more competitive and more profitable. Another success was the breakthrough blending of Lexan and Valox polymers, compounded with suitable fillers, additives, and colors (Xenoy) to meet the special requirements for automotive bumpers.

On the other hand, the managers of GE's diamond business never chose to integrate forward to make diamond wheels, saws, and drills. Such a move would certainly have stimulated the established makers of wheels and saws to buy more synthetic diamond from De Beers.

From time to time grade proliferation had to be reversed, as in a 1994 program in GE engineering plastics, which targeted a 50% reduction. Such a program frees up plant capacity and substantially reduces costs.

Dispose of Low-Return, Low-Growth Businesses

While silicones, diamonds, and engineering plastics were growing rapidly, GE chemical managers chose over time to exit from several marginal product lines that they judged no longer worth continued participation and new investment. Although GE gave up significant sales and some profit, writeoffs were usually not required, because the sale price was greater than the traditionally conservative GE book valuation of plant and inventory assets.

The alkyd resin business was sold after a tragic plant explosion in 1956. Prior years of declining margins reflected increased competition as patent importance declined, technology became widely known, and large paint customers began manufacturing their own required alkyd resins. For these reasons, adequate return on a new plant investment was judged unlikely.

The molded-plastics parts business, the largest department in GE's chemical business in 1945, was sold in 1959. Its unique position during the 1930s and 1940s had eroded as thermoplastics replaced thermosets and as large customers installed their own molding lines. This disposition was a near first within corporate GE.

The small Alnico magnet business, located in Edmore, Michigan, was sold in 1972. This product line had been developed by the Research Lab in the 1920s to produce radio speaker and watt-hour-meter magnets. A later GE technology (Lodex) proved to have limited utility, and GE had no position in some new ceramic magnet technology.

Although Textolite brand decorative laminates had been part of the GE product line from earliest days, market leadership had never been approached. A second plant acquisition in 1974 proved a serious mistake and was resold. GE subsequently sold the entire decorative products line in 1979 to concentrate on its successful industrial laminates.

Purified magnesium oxide (MgO), the insulation medium inside metal-sheathed electric heating elements, had been a successful product since before the chemical division was formed. Over the years several competitors became competent in the technology, were able to enter the field with moderate investment, and competed effectively with declining prices. GE left the business in 1981.

While GE's engineering thermoplastics sales were growing rapidly in the 1970s, the original phenolic resins and molding compounds lines had stagnated, even though GE chemists had added important new technology such as Methylon resins for drum linings and Genal plastics for more rapid mold cycles. The business was sold in 1982 to Plenco, a specialty phenolics producer.

The largest product disposition in this history was Metallurgical Products (originally Carboloy Corporation), which made sintered tungsten carbide tools and inserts for metal cutting tools and for oil drill bits. The business had been a head-to-head U.S. competitor with the Kennametal Corporation for many years, though neither of them was as large worldwide as Sandvik, a Swedish company. After one large oil drill-bit customer succeeded in making its own carbide inserts, Hughes Tool asked GE to sell its Houston plant or face loss of the Hughes business, because this customer was also determined to begin carbide manufacture. GE sold its mining-component carbide business to Hughes Tool.

In the tungsten carbide metal-cutting tool segment the company was unsuccessful in a merger bid for Fagersta, another Swedish producer. After a succession of GE managers subsequently failed to arrest a decline in financial results, the balance of the Detroit-based business was sold to Sandvik in 1986. In addition to giving up the tungsten carbide business, which had once been the largest and most profitable department in the division, the disposition cost GE's lighting business a preferred-supplier position for tungsten metal.

A small industrial paint line (trade name, Glyptal) was sold in the early 1980s, and the Schenectady-based insulation products such as wire enamels, varnishes, mica products, and tapes were sold as a package in 1988.

In sum perhaps a half-billion of today's dollars of product sales were given up by these dispositions. The proceeds, of course, offset some of the large investments continually needed in the rapid-growth product lines. GE managers were confident that the dispositions recovered fair value and that the process helped the division focus on growth products where the company was a continuing leader.

Integrate Backward Selectively

Backward integration, to make in-house a previously purchased material, should gain worthwhile cost reduction and satisfactory investment return, and may yield other benefits. The following initiatives by GE were successful.

In the silicone flow sheet, reuse of byproduct hydrogen chloride to make methyl chloride from methanol was environmentally helpful as well as cost effective. Manufacture of fumed silica powder for silicone rubber reinforcement provided cost improvement and also some byproduct reduction by converting used silicon powder to trichlorosilane and thence to the silica.

At Mt. Vernon and other polycarbonate plants, four backward-integration processes were important additions to the original flow sheet: (1) phosgene manufacture from coke, oxygen, carbon dioxide, and chlorine; (2) BPA from phenol and acetone; (3) brine recycle through electrolytic cells to recover sodium hydroxide and chlorine; and lastly, (4) a world-scale phenol plant using a two-step oxidation of the petrochemical, cumene. The optimized BPA process and the phenol manufacture each yielded purity benefits that were important for making improved PC polymer.

High-impact polystyrene manufacture was added to the Selkirk Noryl plant in a joint venture with the Huntsman Company.

GE did not pursue two Research Lab process discoveries for making plastics monomers: a catalytic oxidation of p-xylene to make terephthalic acid, and a process for making 1,4 butanediol. GE still purchases each of these monomers in large volume to make Valox polyester plastics. A major Corporate R & D program of the late 1980s explored mixtures of low polymers (cyclics) with fiber reinforcements. GE Plastics did not pursue this radical innovation for plastics processing; nor did GE ever try to make elemental silicon.

Organize in Focused Management Units

Since 1945, when the various chemical and semifinished materials businesses were combined in a single division, GE has used many organization forms in managing its chemical activities. The evolution has typically favored small, specially focused management units in preference to larger pooled activities that might be theoretically more cost effective.

The different chemical product lines were managed separately, with strategies that reflected their different technology and market situations. The earliest divisional pooled-sales organization was quickly scrapped in favor of specialized sales forces for the different products. The silicone business became more visible and focused with separate management and department status in 1952, two years before it broke even. The diamond product section had its own manager from the beginning in 1955; and in 1967 when its growth required a new location, it emerged from within Metallurgical Products as a separate department. The Lexan development project was managed as a stand-alone business by the Chemical Development Operation (CDO) for four years, separate from the established phenolic molding compounds business. The later PPO/Noryl project was also managed within CDO and remained independent for about 6 years, even building a $10 million PPO plant at a new location. Although plant investment cost would have been substantially less if this product had been added at the existing Mt. Vernon location, the strong desire for independent, focused management argued for the new site development at Selkirk. Both the product and the site proved to be winners.

Even after Noryl products joined Lexan in the renamed Plastics Department (1968), separate development and marketing organizations competed with each other for business for several years. Internal competition (and some conflict) was encouraged in the belief that each product-line group would be stimulated. When GE did organize a pooled sales force in 1974 to handle engineering plastics sales to molders, the market-development specialists calling on endusers still represented separate products.

GE Plastics managed the European area, on the other hand, on a geographical basis from the beginning. There, substantial management and technical independence from the U.S. organization was encouraged. Similarly, GE Plastics now has a country manager for Japan, but the joint venture companies in Japan are quite self-sufficient technically. Friendly competition and good information-flow between similar plants of the GE Plastics network has proven effective in sharing best practices and product improvements from many sources.

Seek Acquisitions or Joint Ventures with a Good Strategic Fit and Market-Leadership Opportunity

GE didn't need Borg Warner's ABS and related businesses for world leadership in engineering plastics. The explosive 20-year (1968–1988) growth of Lexan, Noryl, Valox, and their derivatives, plus the later Ultem, had already established the company as world leader in this market niche. However, the unexpected opportunity came to grow by acquisition with a good strategic fit.

The strong 1988 plastics market and Borg Warner's leading ABS position brought competitive bidding to the high $2.3 billion winning level. While GE did not achieve all the expected synergies, others, including the Polymerland plastics distribution network and the Cycoloy blend of ABS and PC, have become important business assets. Although ABS plastics per se may be marginally profitable, GE Plastics is now a larger factor in the engineering plastics market, and has higher earnings, for having paid the high price.

Three recent joint ventures illustrate the strategic-fit principle: with BASF to expand production of PBT plastics in Europe; with Bayer AG a joint development company to develop the use of polycarbonate glazing in car windows; and with another division of Bayer AG, a silicone joint venture and capacity increase in Europe.

Company Organization and Culture Contributions

In addition to these chemical management strategies, several aspects of General Electric's organization and culture were crucial or significant contributors to the long-term success. The GE of 1945 may have lacked a management cadre experienced in the integrated chemical polymer and special materials businesses of this history, but it possessed an outstanding industrial Research Laboratory. Without the chemists and chemistry programs of this lab, this business history would not have happened. In addition, many transfers of Research Lab people to the GE chemical businesses have been an important factor in building the latter organization and in maintaining good communications. Works laboratories also contributed by early evaluations of new materials for use in many GE products. The major-appliance plastics lab in particular played a key role in optimizing the early Noryl product line.

Scientists and engineers appreciate an opportunity for open technical contacts. Both Daniel W. Fox and Allan S. Hay, each of whom also had a Du Pont employment offer, were influenced to join GE by the apparent freedom for technical contacts in the company, plus some opportunity to follow projects of their own interest.

As evidenced by its patient support of the Research Lab from its founding in 1900, GE top management believed in technical innovation as a source of growth. Willingness to enter a new product arena if GE could make a technical contribution was part of the company culture. The demands of World War II created several new product and technology opportunities that GE then pursued in the postwar era: aircraft jet engines, land-based gas-turbine-powered generators, radar, flight controls, space programs, and nuclear power are major examples, while silicones is a small example.

Continuous college recruitment and training programs were long established in GE and provided a strong middle-management base for promotions to support expansion. GE has always had a strong management bench. While the bulk of prewar college hires had undergraduate degrees, selective PhD recruiting by the Research Lab, the Works labs, and some engineering groups had established good relations with universities strong in science and engineering. The recruiting programs were expanded in the 1940s and 1950s as GE chemical interests grew, to attract more chemists, chemical engineers, and metallurgists. A "Chemet" training program with multiple work assignments for these graduates paralleled the long-time famous "Test" (see Appendix C) program for electrical and mechanical engineers.

Lastly, GE's talented and well-established accounting, patent, and legal staffs provided quality in-house personnel and experience which were crucially important to fledgling chemical businesses; and, of course, General Electric's size permitted major investment risks to be taken and growth to be financed.

In Chemical Industry Context

GE's commercial development of silicones, synthetic diamond, and engineering plastics, as well as Loctite Corporation's adhesives innovations, are similar to the case histories described by Ralph Landau in his recent chapter, "The Process of Innovation in the Chemical Industry."[3] His two kinds of innovation—the breakthrough, and then the long-sustained improvement in products, processes, and market developments—are demonstrated in each of GE's product successes. Landau notes that the latter innovation processes in the capital-intensive chemical industry often require the financial strength and staying power of large global companies. His examples—polyester fibers and polypropylene plastics—also show the importance of patents and the international character of the producers' investment, sales activity, and competition.

GE's 6 Sigma management project successes (Chapter 12) are similar to those earlier W. Edwards Deming-inspired quality-improvement programs cited by Gordon Cain in *Everybody Wins, A Life in Free Enterprise.*[4]

How Big an Achievement?

The 1998 GE Plastics total sales of $6.6 billion is comparable to the combined Du Pont *Performance Coatings & Polymers* plus *Specialty Polymers*, $8.7 billion sales, or Dow Chemical's *Performance Plastics* segment ($5.1 billion). GE chemical earnings are about equivalent to the combined DuPont segments and exceed those of Dow's *Specialty Polymers.*

As shown below, *Chemical and Engineering News'* 1998 ranking of the top 75 U.S. chemical producers placed GE fourth in both sales and operating profit, after Du Pont, Dow Chemical, and Exxon Chemical.[5]

Top Ten U.S. Chemical Producers–1998

Rank	Chemical Sales ($ mill.)	Chemical Operating Profit ($ mill.)	Operating Profit Margin (%)	Chemical Assets ($ mill.)	Operating Profit/ Assets (%)
1. Du Pont	$26,202	$3,027	11.6	$18,998	15.9
2. Dow Chemical	17,710	2,552	14.4	18,835	13.5
3. Exxon Chemical	10,504	na	*na	na	na
4. General Electric	**6,633**	**1,584**	**23.9**	**9,813**	**16.1**
5. Union Carbide	5,659	803	14.2	7,291	11.0
6. Huntsman	5,200	na	na	na	na
7. ICI Americas	4,900	na	na	na	na
8. Praxair	4,833	856	17.7	8,096	10.6
9. BASF	4,800	na	na	na	na
10. Eastman Chemical	4,481	434	9.7	5,876	7.4

Source: *Chemical and Engineering News,* May 3, 1999.
*Instead of *operating profit,* Exxon reported chemical *net earnings* of $1,213 for 1998 (about $1.9 billion *operating profit*).

We have shown in this history that GE Plastics growth propelled the company's U.S. chemical sales ranking from around 29th in 1968, to 15th in 1984, and to 4th today, a remarkable achievement.

Among world chemical producers for 1997, *Chemical and Engineering News* indicated that GE was 18th in sales, 6th in operating profit, and 1st in profit margin.[6]

(Note: For reference with this "Summation" chapter, the following table, which is taken from annual reports, shows the financial results of GE's chemical business segment from 1984 though 1999.)

In Conclusion

With this summation I rest the case for General Electric's technical and commercial contributions, and well-deserved economic achievement in the chemical industry. It is a history of important scientific innovation, technological achievement, creative market development, and bold investment. People made the difference as successful management strategies were implemented with ever-growing organization competence over the years.

GE Plastics Financial Data from GE Annual Reports

	1984	1985	1986	1987	1988 ($ mill.)	1989	1990	1991
Revenue	$2,241	2,459	2,331	2,751	3,539	4,929	5,167	4,722
Operating Profit	$ 470	466	424	507	733	1,057	1,017	803
Assets*	$2,362	3,876	3,602	3,891	7,130	8,023	7,973	8,340
P&E	$	649	608	380	757	722	693	784
Operating Profit/Revenue	21%	19%	18%	18%	21%	21%	21%	17%
Operating Profit/Assets	16%	14%	12%	13%	**	13%	13%	10%

	1992	1993	1994	1995	1996 ($ mill.)	1997	1998	1999
Revenue	$4,853	5,042	5,681	6,647	6,509	6,695	6,633	6,941
Operating Profit	$ 740	834	981	1,435	1,443	1,500	1,584	1,651
Assets*	$8,081	8,181	8,628	9,095	9,130	8,890	9,813	9,261
P&E	$ 255	376	419	521	748	618	722	462
Operating Profit/Revenue	15%	15%	17%	22%	22%	22%	24%	24%
Operating Profit/Assets	9%	10%	11%	16%	16%	17%	16%	18%

Source: GE Annual Reports, 1984–1999.
*1986–1990 assets include Ladd Petroleum, which was sold in 1990.
**1988 Borg Warner acquisition year (3rd quarter); operating profit/assets not meaningful.

GE Plastics Financial Data from GE Annual Reports

Financial terms notes: GE often recasts previous years financial data slightly to reflect accounting standards changes or product shifts in the company. The data here reflect recasts shown in the 1998 and 1986 annual reports. Other data are as presented in the annual report year.

Operating profit is before income taxes, corporate debt interest, and corporate administrative costs. It is larger than net earnings.

Assets at year-end are those directly associated with the product segment. Product segment assets are greater than the term, invested capital, by the segment liabilities, chiefly accounts payable.

Operating profit/revenue % is one measure of product-segment return on sales. It is greater than net earnings/sales %.

Operating profit/assets % is a different ratio than return on investment % (ROI), but it will track the trends of ROI %.

Appendix: Glossary

A. THERMOPLASTIC POLYMERS, COMPOUNDS, AND BLENDS

ABS	Polymer made from acrylonitrile, butadiene, and styrene; GE Cycolac
ASA	Polymer made from acrylonitrile, styrene, acrylate; GE Geloy
LCP	Liquid crystal polymer
PA	Polyamide, also nylon
PBT	Polybutylene terephthalate; GE Valox
PC	Polycarbonate; GE Lexan
PC/ABS	Blend; GE Cycoloy
PC/ASA	Blend; GE Geloy
PC/PBT	Blend; GE Xenoy
PE	Polyethylene
PEI	Polyetherimide GE Ultem
PEI/PC	Blend; GE Ultem ATX
PET	Polyethylene terephthalate
POM	Polyacetal
PP	Polypropylene
PPO	Polyphenylene oxide; in blend with PS; GE Noryl
PPO/PA	Blend; GE Noryl GTX
PPS	Polyphenylene sulfide
PS	Polystyrene
PSO	Polysulfone
PVC	Polyvinyl chloride

B. Trade Names, Companies, and Chemical Terms

ASEA	Swedish electrical equipment company
ACS	American Chemical Society
Alkanex	GE wire enamel for 130°C service
Alkyd	Resin based on modified phthalic anhydride/glycerin esters
Alnico	Permanent magnet alloy of aluminum, nickel, and cobalt
AKU	Algemene Kunstzijde Unie NV; Dutch chemical company
Bakelite	Union Carbide or generic phenol/formaldehyde plastic
Bayblend	Bayer PC/ABS thermoplastic blend
BASF	Badische Aniline und Soda Fabrik; German chemical company
Bexloy	Du Pont, PA/XX thermoplastic blend
Borazon	GE superabrasive, cubic boron nitride
BOZ	Bergen op Zoom, The Netherlands; GE chemical plant site
BPA	Bisphenol A, 2,2-bis-(4-hydroxyphenyl)-propane
BW	Borg Warner Corporation
BP	British Petroleum Company
Carboloy	GE tungsten carbide tools (or Corporation)
CD	Compact disc
CD-ROM	Compact disc – read-only-memory
CDO	Chemical Development Operation (GE)
CEO	Chief Executive Officer
Celron	Celanese polyacetal thermoplastic
CO	Carbon monoxide
Compax	GE bonded polycrystalline diamond compact shapes
C R&D	Corporate Research and Development Center (GE)
Cycolac	GE ABS thermoplastic
Cycoloy	GE PC/ABS blend thermoplastic
D unit	see methyl silicone polymer units
DC	Dow Corning Corporation
Delrin	DuPont polyacetal thermoplastic
Dulux	DuPont alkyd resin-based paints
DVD	Diamond vapor-deposition process
DVD	Digital video disc
EPL	Engineering Polymers, Ltd.; GE/Nagase Japan joint venture
FDA	Food and Drug Administration
Flamenol	GE wire covered with PVC extruded insulation
Formex	GE wire enamel
Formvar	Resin in GE Formex

FR-4	GE copper-clad epoxy/glass laminates
GELON	GE polyamide (PA) thermoplastic
GELOY	GE ASA thermoplastic
GEM Polymers	GE/Mitsui Petrochemicals joint venture for xylenol and PPO polymer
GENAL	GE fast-molding phenolic compound
GEP	GE Plastics
GETEC	GE PPO/epoxy glass laminate
Glyptal	GE alkyd resin-based industrial paints
Grignard	Chemical reaction, in which an organic halide is first reacted with magnesium in ethyl ether; the Grignard reagent formed can then be reacted with a metallic halide to make another organometallic halide plus (discarded) magnesium halide
Hardcoat	GE silicone resin coating for polycarbonate and other plastics
HCl	Hydrogen chloride
ICI	Imperial Chemical Industries
IGE	International General Electric Company
IMD	Insulating Materials Department (GE)
ITT	International Telephone and Telegraph Company
IUPAC	International Union of Pure and Applied Chemistry
Isonel	GE high-temperature wire enamel
JACS	*Journal of the American Chemical Society*
KOH	Potassium hydroxide
Lexan	GE polycarbonate (PC) thermoplastics
Loctite	Fast-curing anaerobic adhesive (or Corporation)
Lomod	GE thermoplastic elastomers
M unit	see methyl silicone polymer units
MBS	GE diamond for metal-bonded saws
MBG	GE diamond for metal-bonded grinding wheels
Megadiamond	Bonded polycrystalline compact developed by Tracy Hall
Methyl silicone polymer units:	
M unit	Monofunctional, chain-terminating, $(CH_3)_3SiO_{1/2}$
D unit	Difunctional, linear chain, $(CH_3)_2SiO_{2/2}$
T unit	Trifunctional, cross-linking, $CH_3SiO_{3/2}$
Q unit	Quatrofunctional, cross-linking, $SiO_{4/2}$
Methylon	GE phenolic-based drum-lining resin
MPD	Metallurgical Products Department (GE)
MgO	Purified magnesium oxide powder, an insulator at red heat

Micamat	GE mica paper, usually bonded
Mypolex	Du Pont diamond dust for polishing diamonds
Neoprene	Du Pont chlorinated organic synthetic rubber
Noryl	GE modified-PPO thermoplastic
Noryl GTX	GE PPO/PA blend thermoplastic
NSERC	Natural Sciences and Engineering Research Council of Canada
NSR	GE nitrile-modified silicone rubber
Nylon	Fiber or thermoplastic polymer with polyamide linkages
OEM	Original equipment manufacturer
Pan-Glaze	Dow Corning silicone resin coating for bread pans
P&L	Profit and loss
PCBs	Polychlorinated biphenyls
Permafil	GE impregnating varnish
Phenolics	Resins or molding compounds derived from polymers made by reacting phenol and formaldehyde
Pocan	Bayer PBT thermoplastic
Praxair	Liquefied and compressed gases company, formerly Linde Division of Union Carbide
Prevex	Borg Warner polyphenylene ether thermoplastic
PSA	Pressure-sensitive adhesive
PUFA	Polyurethane foam additive
Pyrrhic	Worthless, outcome not worth the cost
Q unit	see methyl silicone polymer units
R&D	Research and development
RF	Radio frequency
RIM	Resins and Insulating Materials Dept. (GE)
RTV	GE room-temperature vulcanizing silicone rubber
RVG	GE diamond for resin-bonded grinding wheels
6 Sigma	Project management technique for greatly improving the quality control and reliability of corporate processes. Sigma means the "standard deviation" of a normal distribution curve around a center point. 6 Sigma control means a process having 6 standard deviations between the center point and the upper and lower spec limits. Reference 5 of Chapter 12 describes the 6 Sigma theory, including the research-supported assumption that most processes drift 1.5 sigma from the original center point
Silastic	Dow Corning silicone rubber compound
Sil-Plus	GE silicone reinforced rubber gum base

SPE	Society of Plastics Engineers
Supec	GE/Toso Susteel polyphenylene sulfide thermoplastic
Stratapax	GE bonded polycrystalline compact for oil drill-bits
T unit	see methyl silicone polymer units
Textolite	GE decorative laminate
TIPCO	India plastics distributor; GE joint venture partner
Tosil	GE–Toshiba silicones joint venture
UC	Union Carbide Corporation
UK	United Kingdom
Ultem	GE polyetherimide (PEI) thermoplastic
UV	Ultraviolet light
Valox	GE polybutylene terephthalate (PBT) thermoplastic
Xenoy	GE PC/PBT blend thermoplastic
Zytel	DuPont polyamide (PA) thermoplastics

C. GE ORGANIZATION NOTES

GE Organization units in order of increasing size (beginning around 1950):

Sections, within a Department or Operation:
Engineering (later renamed Research and Development in chemical departments)
Manufacturing
Marketing (Sales function included within Marketing)
Finance
Employee and Community Relations
Legal

Operation   P&L unit without all the functions.
Department   Smallest full-function P&L unit. Headed by a General Manager.
Division   Comprises several related departments. Sometimes included a pooled sales force for several product lines. Headed by a Vice-President and General Manager.
Group   Comprises related divisions. Sometimes included a pooled sales force. Headed by Senior Vice-President and Group Executive.
Sector (1976–1985) Several related Groups. Headed by an Executive Vice-President and Sector Executive.

(Note: These rigorous organization descriptors disappeared from the GE lexicon after 1986, part of the CEO's drive toward the "boundaryless corporation.")

The GE "Test" program (later EMP "Engineering Management Program") for entry-level electrical and mechanical engineers comprised multiple three-month rotating assignments at major plants. The name derived from the work assignments, which were the testing of completed GE electrical products.

Organization and nomenclature evolution for GE's chemical business:

1945   Chemical Department
1948   Chemical Division
1951   Chemical and Metallurgical Division
1968   Chemical and Medical Division +
        Industry Components and Metallurgical Division
1972   Chemical and Metallurgical Division
1973   Chemical and Metallurgical Division +
        Plastics Division
1977   Engineered Materials Group
1982   Engineered Materials Group +
        Plastics Operations
1983   Engineered Materials Group +
        Plastics Group
1986   Plastics
1987   GE Plastics

Chapter References

Chapter 1 References: What's General Electric doing in the chemical business?

1. Peters, T. J., and R. H. Waterman, Jr., *In Search of Excellence, Lessons from America's Best-Run Companies*, Harper & Row, New York, (1982).
2. *Business Week*, Mar. 24, 1997.
3. *General Electric Annual Reports*, 1948,1968,1984,1997, 1998.
4. *Chemical Week*, May 28, 1997.
5. *Chemical and Engineering News*, June 16, 1969; May 6, 1985; May 3, 1993, May 9, 1994; May 8, 1995; May 6, 1996; May 5, 1997; May 4, 1998; May 3, 1999.
6. U.S. Department of Commerce, "The National Medal of Technology Nomination Criteria"; President George Bush, Rose Garden, Sept. 16, 1991.

Chapter 2 References: Early years of GE Chemistry: 1900–1948

In addition to the references cited, inputs to this chapter were provided by Donald Brown, Wyman Goss, James Pyle, Charles Reed, and George Wise.

1. Wise, G., *Willis R. Whitney, General Electric, and the Origins of U.S. Industrial Research*, Columbia Univ. Press, New York, (1985).
2. Wise, G., private communication to J. T. Coe, 1996.
3. Liebhafsky, H. A., *Silicones Under the Monogram, A Story of Industrial Research*, Wiley, New York, (1978).
4. General Electric Organization Lecture Series, 1940–1941, Given to Test and BTC Students, Plastics Department.
5. "One Plastics Avenue," *Modern Plastics*, (1938).

6. *Eugene G. Rochow*, interview conducted by James J. Bohning, at Ft. Myers, FL, January 24, 1995 (Philadelphia: Chemical Heritage Foundation, Oral History No. 129).

7. *Charles Reed*, interview conducted by Leonard Fine and George Wise, at Columbia University, July 11, 1986 (Philadelphia: Chemical Heritage Foundation, Oral History No. 51).

8. Reed, C. E., and J. T. Coe, U.S. Patent No. 2,389,931 (Nov. 27, 1945).

9. Schubert, A. E., and C. E. Reed, U.S. Patent No. 2,563,557 (Aug. 7, 1951).

10. Patnode, W. I., U.S. Patent Nos. 2,469,888 and 2,469,890 (May 10, 1949).

11. Warrick, E. L., *Forty Years of Firsts, the Recollections of a Dow Corning Pioneer*, McGraw-Hill, New York, (1990).

12. *J. Franklin Hyde*, interview conducted by James J. Bohning, at Marcos Island, FL, on April 30, 1986 (Philadelphia: Chemical Heritage Foundation, Oral History No. 26).

13. *Earl L. Warwick*, interview conducted by James J. Bohning, at Midland, MI, January 16, 1986 (Philadelphia: Chemical Heritage Foundation, Oral History No. 45).

Chapter 3 References: GE Silicones: 1940–1964

In addition to the references cited, inputs to this chapter were provided by Donald Brown, Joseph Caprino, Hart Lichtenwalner, Charles Reed, John Stauffer, and Ronald Wishart.

1. Warrick, E. L., *Forty Years of Firsts, The Recollections of a Dow Corning Pioneer*, McGraw-Hill, New York, (1990).

2. MacGregor, R. R., *Silicones and their Uses*, McGraw-Hill, New York, (1954).

3. *J. Franklin Hyde*, interview conducted by James J. Bohning, at Marcos Island, FL, on April 30, 1986 (Philadelphia: Chemical Heritage Foundation, Oral History No. 26).

4. *Earl L. Warwick*, interview conducted by James J. Bohning, at Midland, MI, January 16, 1986 (Philadelphia: Chemical Heritage Foundation, Oral History No. 41).

5. Liebhafsky, H. A., *Silicones Under the Monogram, A Story of Industrial Research*, Wiley, New York, (1978).

6. Cordiner, R. J., *New Frontiers for Professional Managers*, McGraw-Hill, New York, (1956).

7. Meals, R. N., and F. M. Lewis, *Silicones*, Reinhold, New York, (1959).

8. Polmanteer, K. E., "Silicone Rubber, Its Development and Technological

Progress," at ACS Rubber Div. Meeting, Cleveland, OH, (1987).

9. Coe, J. T., "Measuring the Success of R&D Innovations," *National Productivity Review,* (1982).

10. Reed, C. E., "Old Wisdom and New Businesses," General Electric Management Conference, Belleair, FL, (1979), (unpublished).

11. *Daniel W. Fox,* interview conducted by Leonard Fine and George Wise, at Pittsfield, MA, August 14, 1986 (Philadelphia: Chemical Heritage Foundation, Oral History No. 58).

Chapter 4 References: Loctite

1. Krieble, R. H., "Anaerobic Adhesives–A Solution that Found a Problem," *Research Management,* XII, (1980).

2. "Adhesives–Loctite Corp.," *Successful Product and Business Development,* N. H. Giragosian, Ed., Commercial Development Association, Dekker, New York, (1978).

3. *Chemical and Engineering News,* Dec. 9, 1996.

Chapter 5 References: Synthetic diamond

In addition to the references cited, inputs to this chapter were provided by Harold Bovenkerk, William Cordier, Louis Kapernaros, John Kennedy, and George Wise. Wise is author of the diamond history portion.

1. Hannay, N. B., *Proc. Roy. Soc.,* 30, 188 (1880).

2. Bannister, F. A., and K. Lonsdale, *Nature,* 151, 334 (1943).

3. Lonsdale, K., *Nature,* 196, 104 (1962).

4. Moissan, H., *C. R. Acad. Sci.,* 118, 320 (1901).

5. Parsons, C. A., *Phil. Trans. A.,* 220, 67 (1920).

6. Desch, C. H., *Nature,* 121, 799 (1928).

7. De Vries, R. J., *J. Mater. Educ.,* 13, 387, (1991).

8. Rossini, F. D., and R. S., Jessup, *J. Res. Nat. Bur. Stand.,* 21, 491 (1938).

9. Bridgman, P. W., *J. Appl. Phys.,* 12, 461 (1941).

10. Liander, H., *Ind. Diamond Rev.,* 412 (1980).

11. Bridgman, P. W., *J. Chem. Phys.* 15, 92 (1947).

12. Mogerman, W. D., *Zay Jeffries,* American Society for Metals, Metals Park, OH, (1973).

13. Bridgman, P. W., *Scientific American,* 193, 42 (1955).

14. Angus, J. C., "Development of Low Pressure Diamond Growth in the United States," *Synthetic Diamond,* Spear and Dismukes, eds., (1994).

15. De Vries, R. J., *Ann. Rev. Mater. Sci.,* 17, 161 (1987).

16. Suits, C. G., "The Synthesis of Diamond–A Case History in Modern Science," ACS Meeting, Rochester, NY, (1960).

17. GE Engineering Council Minutes (unpubl.) (1951).
18. Oriani, R. A., and W. A. Rocco, GE Memo MA-36, (1957).
19. Strong, H. M., Laboratory notebook (1953).
20. Hall, H. T., Laboratory notebooks (1953, 1954).
21. Bundy, F. P., H. T. Hall, H. M. Strong, and R. H. Wentorf, Jr., *Nature*, 176, 51 (1955).
22. Bundy, F. P., *J. Appl. Phys.*, 38, 631 (1964).
23. Strong, H. M., *J. Phys. Chem.*, 75, 1838 (1971).
24. Hazen, R. M., *The New Alchemists*, Times Books, New York, (1993).
25. Reed, C. E., "Old Wisdom and New Businesses," General Electric Management Conference, Belleair, FL, (1979).
26. Strong, H. M., U.S. Patent Nos. 2,947,609 (Aug. 2, 1960).
27. Hall, H. T., H. M. Strong, and R. H. Wentorf, Jr., U.S. Patent No. 2,947,610 (Aug. 2, 1960).
28. Hall, H. T., U.S. Patent Nos. 2,941,248 (June 21, 1960).
29. Bovenkerk, H. P., F. P. Bundy, H. T. Hall, H. M. Strong, and R. H. Wentorf, Jr., "Preparation of Diamond," *Nature*, 184 (1959).
30. Hall, H. T., "Ultra-High-Pressure, High Temperature Apparatus: The 'Belt'", *Rev. Sci. Instrum.*, 31 (2), (1960).
31. Wentorf, R. H., Jr., "High Pressures and Synthetic Diamonds," XVIIth Int. Congress of Pure and Applied Chemistry, Munich, Germany, (1960).
32. Bundy, F. P., H. P. Bovenkerk, H. M. Strong, and R. H. Wentorf, Jr., "Diamond-Graphite Equilibrium Line from Growth and Graphitization of Diamond," *J. Chem. Phys.*, 35 (2) (1961).
33. Wentorf, R. H., Jr., "Synthesis of the Cubic Form of Boron Nitride," *J. Chem. Phys.*, 34, (3) (1961).
34. Davis, R. F., Editor, *Diamond Films and Coatings*, Noyes Press, Park Ridge, NJ, (1993).
35. "GE Crushes the Trustbusters," *American Lawyer*, Jan./Feb. 1995.

Chapter 6 References: Lexan polycarbonate: 1953–1968

In addition to the references cited, inputs to this chapter were provided by William Christopher, Walter Dugan, Robert Hatch, George McCullough, Leroy Moody, Charles Reed, and George Wise.
1. *Daniel W. Fox*, interview conducted by Leonard Fine and George Wise, at Pittsfield, MA, August 14, 1986 (Philadelphia: Chemical Heritage Foundation, Oral History No. 58).
2. Fox, D. W., "History and Commercial Development of Polycarbonates," Address to the Society of Plastics Engineers,
3. Fox, D. W., "From Plant Pigments to Polymers," *The Chemist*, Sept. 1987.

4. Morone, J., "New Business Development and General Managerial Decision Making," *Research on Technological Innovation, Management and Policy*, Vol. 5, JAI Press, (1993).

5. Christopher, W. F., and D. W. Fox, *Polycarbonates*, Reinhold, New York, (1962).

6. Schnell, H., "Polycarbonates, a Group of New Types of Thermoplasts: Preparation and Properties of Aromatic Esters of Carbonic Acid," *Angew. Chem.*, 68, (1956).

7. *Charles Reed*, interview conducted by Leonard Fine and George Wise, at Columbia University, July 11, 1986 (Philadelphia: Chemical Heritage Foundation, Oral History No. 51).

8. Christopher, W. F., "Lexan (r) Polycarbonate Resin, a New High Strength, Heat Resistant Plastic," Society of Plastics Engineers Technical Conference, Detroit, MI (1958).

9. *Chemical Week*, June 6, 1959; July 11, 1959.

10. Fine, L. W., *Chemistry for Nonchemists–A Practical Guide to the Science and Technology of GE Plastics*, Columbia University, (1980) (unpublished).

11. Schnell, H., *Chemistry and Physics of Polycarbonates*, Interscience, New York, (1964).

12. Forrestal, D. J., *Faith, Hope, & $5,000, the Story of Monsanto*, Simon & Shuster, New York (1977).

Chapter 7 References: Noryl thermoplastic: 1956–1968

In addition to the references cited, inputs to this chapter were provided by Reuben Gutoff, Stephen Hamilton, Allan Hay, Leroy Moody, Charles Reed, William Vivian, Jack Welch, and George Wise.

1. *Allan S. Hay*, interview conducted by Leonard Fine and George Wise, at Schenectady, NY, July 24, 1986 (Philadelphia: Chemical Heritage Foundation, Oral History No. 57).

2. "The Process of Innovation in the Chemical Industry," R. Landau, *Chemicals and Long-Term Economic Growth, Insights from the Chemical Industry*, Chap. 5, A. Arora, R. Landau, and N. Rosenberg, eds. Wiley, in conjunction with the Chemical Heritage Foundation, New York, (1998).

3. Hay, A. S., H. S. Blanchard, G. F. Endres and J. W. Eustance, "Polymerization by Oxidative Coupling," *J. Amer. Chem. Soc.*, 81 (1959).

4. Hay, A. S., "Polymerization by Oxidative Coupling: Discovery and Commercialization of PPO and Noryl Resins," *Journal of Polymer Science: Part A: Polymer Chemistry*, Vol. 36, Wiley, New York (1998).

5. *Daniel W. Fox*, interview conducted by Leonard Fine and George Wise, at Pittsfield, MA, August 14, 1986 (Philadelphia: Chemical Heritage Foundation, Oral History No. 58).
6. *Charles Reed*, interview conducted by Leonard Fine and George Wise, at Columbia University, July 11, 1986 (Philadelphia: Chemical Heritage Foundation, Oral History No. 51).
7. *GE Monogram*, 1988.
8. "Noryl Resins–General Electric Company," *Successful Product and Business Development*, N. H. Giragosian, ed., Commercial Development Association, Dekker, New York (1978).
9. Hamilton, S. B. U.S. Patent No. 3,446,856 (1969).
10. Morone, J., "New Business Development and General Managerial Decision-Making," *Research on Technological Innovation, Management and Policy*, 5 (1993).
11. Boldebuck, E. M., U.S. Patent No. 3,063,872 (1962).
12. Cizek, E. P., U. S. Patent No. 3,383,435 (May 14, 1968).
13. Fine, L. W., *Chemistry for Nonchemists–A Practical Guide to the Science and Technology of GE Plastics*, Columbia University, New York (1980) (unpublished).

Chapter 8 References: GE engineering plastics: 1968–1987
In addition to the references noted below, inputs to this chapter were provided by Paul Dawson, Albert Gilbert, Philip Gross, Allan Hay, Jean Heuschen, Charles Reed, Jack Welch, and Joseph Wirth.
1. *High Performance Polymers: Their Origin and Development*, Symposium, Elsevier, Amsterdam (1986): "Engineering Plastics: The Concept that Launched an Industry," L. H. Gillespie (Du Pont); "A Path to ABS Thermoplastics," W. A. Pavelich (Borg Warner Chemicals); "History, Aromatic Polycarbonates," D. W. Fox (General Electric); "The History of Poly(Butylene Terephthalate) Molding Resins," D. McNally and J. S. Gall (Celanese); "Xenoy and Noryl GTX Engineering Thermoplastics Blends," J. M. Heuschen (General Electric Plastics Europe); "Discovery and Development of Polyetherimides," J. G. Wirth (General Electric).
2. Society of the Plastics Industry (SPI) Statistical Dept., from Tariff Commission Report S.O.C. Series P-68, for 1968.
3. J. F. Welch interview by J. T. Coe, Feb. 14, 1997.
4. *Daniel W. Fox*, interview conducted by Leonard Fine and George Wise, at Pittsfield, MA, August 14, 1986 (Philadelphia: Chemical Heritage Foundation, Oral History No. 58).

5. *Charles Reed*, interview conducted by Leonard Fine and George Wise, at Columbia University, July 11, 1986 (Philadelphia: Chemical Heritage Foundation, Oral History No. 51).

6. Morone, J., "New Business Development and General Managerial Decision-Making, *Res. Technol. Innovation, Manage. Policy*, 5 (1993).

7. J. D. Opie, interview by J. T. Coe, Oct. 21, 1997.

8. Fine, L. W., *"Chemistry for Nonchemists–A Practical Guide to the Science and Technology of GE Plastics,"* Columbia University, New York (1980) (unpublished).

9. R. H. Jones, interview by J. T. Coe, Oct. 10, 1997.

10. *Allan S. Hay*, interview conducted by Leonard Fine and George Wise, at Schenectady, NY, July 24, 1986 (Philadelphia: Chemical Heritage Foundation, Oral History No. 57).

11. G. H. Hiner, interview by J. T. Coe, Feb. 11, 1998, and several telephone interviews in 1998.

12. *Chemical and Engineering News*, Feb. 22, 1982, p.23.

13. Smith, E.M., "How GE Plastics Became an International Business," a log prepared by Eva M. Smith, last update May 1995 (unpublished).

14. J. M. Heuschen, interview by J. T. Coe, Dec. 6, 1997.

Chapter 9 References: Growth by means of a major acquisition: 1988–1991

In addition to the references cited, inputs to this chapter were provided by Paul Dawson, Philip Gross, Robert Hess, Glen Hiner, Robert Muir, Jack Peiffer, and Joseph Wirth.

1. Cordiner, R. J., *New Frontiers for Professional Managers*, McGraw-Hill, New York (1956).

2. *Wall Street Journal*, June 17, 1988.

3. *Chemical Week*, June 22, 1988.

4. *GE 1988 Annual Report.*

5. *New York Times*, Nov. 15, 1995.

6. G. L. Rogers, interview by J. T. Coe, Mar. 15, 1996.

7. J. F. Welch, interview by J. T. Coe, Feb. 14, 1997.

8. J. Hueschen, interview by J. T. Coe, Mar. 2, 1998.

Chapter 10 References: Laminates and insulation products

Inputs to this chapter were provided by Michael Abrams, Swede Berntson, Dean Daniels, Wyman Goss, John Loritsch, Theodore Ohart, Frank O'Keefe and James Pyle.

Chapter 11 References: GE Silicones: 1965–1998

In addition to the references cited, inputs to this chapter were provided by Richard Brown, Joseph Caprino, Wayne Delker, Thomas Gensler, Philip Gross, Philip Jeszke, Thad Leister, Hart Lichtenwalner, Paul McBride, Donald H. Miller, Richard Moeller, and Jeff Welde.

1. Warrick, E. L., *Forty Years of Firsts, The Recollections of a Dow Corning Pioneer*, McGraw-Hill, New York (1990).
2. Liebhafsky, H. A., *Silicones Under the Monogram, A Story of Industrial Research*, Wiley, New York (1978).
3. *Chemical Week*, May 28, 1997.

Chapter 12 References: Engineering plastics: 1992–1998

In addition to the references cited, inputs to this chapter were provided by Robert Hess, Jean Heuschen, Robert Muir, and Gary Rogers.

1. GE Annual Reports, 1990 through 1997, inclusive.
2. *Chemical and Engineering News*, Dec. 21, 1998.
3. *Chemical Week*, May 28, 1997.
4. Heuschen, J., "Maximizing Technology: Strategies for Becoming a World Class Supplier," Society of the Plastics Industry Meeting, Lake Geneva, WI (1996).
5. Harry, M. J., *The Vision of Six Sigma: A Roadmap for Breakthrough*, Sigma, Phoenix, AZ (1994).

Chapter 13 References: People make the difference

In addition to the references cited, inputs to this chapter were provided by Francis Bundy, David Hall, Herbert Strong, and Rolf Wentorf.

1. *Eugene G. Rochow*, interview conducted by James J. Bohning in Ft. Myers, FL, January 24, 1995 (Philadelphia: Chemical Heritage Foundation, Oral History No.129).
2. *Daniel W. Fox*, interview conducted by Leonard Fine and George Wise, at Pittsfield, MA, August 14, 1986 (Philadelphia: Chemical Heritage Foundation, Oral History No. 58).
3. *Allan S. Hay*, interview conducted by Leonard Fine and George Wise at Schenectady, NY, July 24, 1986 (Philadelphia: Chemical Heritage Foundation, Oral History No. 57).
4. Fox, D. W., letter to C. E. Reed, Mar. 21, 1988.
5. Fine, L. W., *Chemistry for Nonchemists–A Practical Guide to the Science and Technology of GE Plastics*, Columbia University, (1980) (unpublished).

6. Liebhafsky, H. A., *Silicones Under the Monogram, A Story of Industrial Research*, Wiley, New York (1978).

7. U. S. Department of Commerce, *The National Medal of Technology - Nomination Criteria*; President George Bush, Sept. 16, 1991.

8. Reed, C. E., "Old Wisdom and New Businesses," General Electric Management Conference, Belleair, FL (1979) (unpublished).

9. Morone, J., "New Business Development and General Managerial Decision-Making," *Res. on Technol. Innovation, Manage. Policy*, 5 (1993).

10. R. H. Jones, interview by J. T. Coe, Oct. 10, 1997.

11. J. F. Welch, interview by J. T. Coe, May 15, 1998.

12. G. H. Hiner, interview by J. T. Coe, Feb. 11, 1998; also several telephone interviews in 1998.

13. G. S. Rogers, interviews by J. T. Coe, Mar. 15, 1996 and Mar. 2, 1998.

14. *Chemical Week*, May 28, 1997.

15. GE Annual Reports, 1992 through 1997.

16. *Chemical and Engineering News*, May 6, 1996, May 5, 1997, May 4, 1998, July 20, 1998.

Chapter 14: Summation

1. GE 1998 Annual Report.

2. *Chemical Week*, May 28, 1997.

3. Landau, R., "The Process of Innovation in the Chemical Industry," *Chemicals and Long-Term Economic Growth, Insights from the Chemical Industry*, Chap. 5, A. Arora, R. Landau, and N. Rosenberg, eds., Wiley, in conjunction with the Chemical Heritage Foundation, New York (1998).

4. Cain, G., *Everybody Wins, A Life in Free Enterprise*, Chemical Heritage Foundation, New York (1997).

5. *Chemical and Engineering News*, July, 20, 1998.

Names Index

(boldface indicates photograph)

Abbott, Joseph, 41
Abrams, Michael J., 145
Agens, Maynard C., 18
Anderson, R. P., 88
Angus, John C., 52
Arrhenius, Svante, 51
Baekeland, Leo H., 9-11, 15
Barr, Kenneth, 94
Barry, Arthur J., 28
Bass, Shailer L., 28
Beardsley, W. Kenneth, 60
Becalli, Ferdinando F. 145
Berridge, Charles A., 40
Blessing, Olin D., 28
Bob and Ray, 93
Boldebuck, Edith, 88
Boman, Pim, 96
Borch, Fred J., 78, 102, 172
Bossidy, Lawrence A., 116
Bovenkerk, Harold P., 53, 54, 58, 62, 162
Boyd, Austen W., 35, 40
Bridgman, Percy W., 50-53
Bueche, Arthur M., 103
Bundy, Francis P., 53, **54,** 55-56, 162-163
Bunger, Frederick, 120
Burnett, Robert E., 45-48
Bush, President George H. W., **8,** 170
Busink, Jan, 94
Cain, Gordon, 190
Caprino, Joseph C., 132
Carnegie, Andrew, 1
Carothers, Wallace H., 15
Carson, Charles R., 106
Cass, William E., 73
Castles, John T., 78, 91
Cheney, James E., 53, **54,** 58, 162
Christopher, William F., 74-76, 164
Cizek, Eric P., 88
Clausius, Rudolph, 51
Coe, Jerome T., 18, 37, **39,** 41, 43
Coes, Loring, Jr., 53
Collings, William R., 28, 30-31
Cook, Newell, 103
Coolidge, William D., 13-14, 17, 24, 168
Cordier, William K., 58, 61-63, 65
Cordiner, Ralph J., 23, 30, 33, 44,

56, 57, 76, 78, 113, 126
Crew, Charles E., 146
Daily, Robert T., 131
D'Alelio, Frank, 12
Davis, John L., 18
Davy, Humphry, 49, 50
Dawson, Paul L., 116
Debacher, Donald E., 101, 103
De Carli, Paul S., 66
Deming, W. Edwards, 190
Dotson, James M., 40
Dow, Willard H., 31
Driscoll, William P., 145
Dugan, Walter J., 78
Edison, Thomas, 9
Edwards, Robert, 120
Elliott, John R., **169**
Espe, Matthew J., 145
Eversole, W. G., 52, 66
Faraday, Michael, 50
Fenn, Howard N., 28
Fiedler, E. F., 73
Fines, Robert, 145
Finholt, Robert, 88
Fisher, John C., 78
Fitzgerald, Thomas H., 131, 132
Flynn, Edward J., 14, 18
Foss, Peter N., 145
Fox, Daniel W., 69, 71, **72**, 73-77, 80, 85, 88, 93, 95-96, 98, 103, 112, 120, 128, 164, **165**, 166, 170, 189
Giaever, Ivar, 10
Gibbs, J. Willard, 51
Gibson, Bob, 93
Gibson, Robert L., 43-44, 57, 60-61, 77-78, 126
Gifford, A. McKay, 12
Gilbert, Albert L., 103
Gillespie, J. Stokes, 58

Gilliam, William F., 18, **19**, 28
Goldberg, Eugene P., 73, 75
Goldblum, Kenneth B., 73
Grosman, Doron, 145
Gutoff, Reuben, 85, **86**, 87-88, 91, 95, 101
Hall, H. Tracy, 53, **54**, 55-56, 65, 68, 162, **163**, 165
Hamilton, Stephen B., 85, 87
Hannay, James B., 50
Harper, Arthur H., 146
Hatch, Robert L., 77
Hay, Allan S., 83-85, **86**, 87, 89, 98, 103, 112, 166, **167**, 189
Hazen, R. M., 56
Heuschen, Jean M., 112, 142, 146
Hewitt, Wayne, 146
Hiner, Glen H., 105-106, 109-110, 116, 118, 139, 174, **175**
Hughes, D., 146
Hunter, Melvin J., 28
Hyde, J. Franklin, 9, 15, 20, 28
Ireland, James W., 146
Jamieson, John C., 66
Jeffries, Zay, 20-21, 23, 31, 51-52
Jessup, Ralph S., 51
Jones, Reginald H., **102**, 103, 172
Kapernaros, Louis, 65
Kauppi, Toivo A., 28
Kennedy, John D., 60
Kienle, Roy H., 11, 13-15
Kipping, Frederic S., 15
Kistler, Samuel, 53
Koscher, Edward, R., 116
Krieble, Robert H., 46-48, 74
Krieble, Vernon K., 46
Kunze, Robert, 94
Landau, Ralph, 190
Langmuir, Irving, 10
Leibhafsky, Herman A., 168

Leipunskii, 51
Liander, Halvard, 56
Lichtenwalner, Hart K., 40
Lidstone, John, 103
Lindblad, Eric, 52
Loncrini, D. F. , 73
Lonsdale, Kathleen, 50
Lucas, Glennard, 32
MacGregor, Rob Roy, 28
Marsden, James, 18
Marshall, Abraham L., 14-17, **19**, 21, 23-24, 53, 69, 71, 160, 162, 168, **169**, 170
McClain, Denny, 93
McCullough, George E., 73, 77
Meals, Robert N., 18
Modic, Frank, 135
Moissan, Henri, 50
Moody, Leroy S., 87
Muir, Robert E., 146
Navias, Louis, 53, 55
Nerad, Anthony J., 53, **54**, 162
Nordlander, Birger W., 45-46, 48
Opie, John D., 98, 172
Oriani, Richard E., 53
Ostwald, Wilhelm, 51
Parsons, Charles A., 50
Patnode, Winton I., 14-15, 18, **19**, 24, 27, 31, 134, 160, 168-169
Pechukas, Alphonse, 71, 164
Peters, Edward, 165
Peters, Tom, 1
Precopio, Frank M., 71
Pyle, James J., 12
Rammrath, Herbert. G., 116
Raynolds, James W., 31-33
Reed, Charles E., 7, **8**, 17-18, 22, 24, 31-33, **39**, 61-63, 74, 77, 85, 87-88, 90-91, 96, 101, 159, 169-170, **171**, 172

Relles, Howard, 103
Rickover, Hyman, 20
Rochow, Eugene G., 15-18, **19**, 21, 23-24, 27-28, **39**, 42, 80, 160, **161**, 168-170
Rogers, Gary L., 118, 139-140, 145-147, 176, **177**
Rossini, Frederick D., 51
Safford, Moyer M., 14
Sato, Hideo, 95, 110
Sauer, Robert O., **19**
Scheiber, W. J., **19**
Schnell, Hermann, 73-74, 164
Schubert, A. Eugene, 74, 76, 78
Sellers, Jesse E., 18
Seral, John M., 146
Shenian, Popkin, 88, 165
Sherwood, Thomas K., 17
Shill, G. Harry, 12
Simpson, L. Donald, 116
Smiddy, Harold F., 30-31
Solmssen, Peter Y., 146
Speier, John, 131
Sprung, M. M., **19**
Steinmetz, Charles P., 10
Stock, Alfred, 16
Strong, Herbert M., 53, **54**, 55-56, 63-64, 162
Suits, C. Guy, 17, 24, 53, 162, 168
Sullivan, Eugene C., 15, 28, 30
Swope, Gerard, 20
Takekoshi, Tohru, 103
Tennant, Smithson, 49
Thompson, Richard J., 75
Turnbull, David, 53
van t'Hoff, Jacobus Henricus, 51
von Platen, Balthazar, 52,
Walsh, M., 146
Warrick, Earl L., 28-29
Wascher, Uwe S., 116, 145

Waterman, Robert H., 1
Webb, James, 103
Welch, John F., Jr., 2, 85, **86**, 87-89, 91, 93-96, 98, 101, **102**, 103, 105, 110, 114-116, 118, 120, 147, 164, 172, **173**, 174
Welsh, Charles, 18
Wentorf, Robert H., 53, **54**, 55-56, 63-65, 162-163

Whitney, Willis R., 10, 13
Williams, Frank, 103
Wilson, Charles E., 20, 33
Wirth, Joseph G., 103, 105, 116, 131, 166
Wise, George, 54
Woodburn, William A., 145
Wright, J. Gilbert E., 14, 18

Subject Index

1,4-butanediol, Valox monomer, 101

2,6-xylenol, 84-85, 166, 175

6 Sigma, 139, 147, 190

ABS resins, 115-118, 141, 188

Alkanex, 69, 71, 126

Alkyd resins, 11, 13, 15, 22-24, 43

Allied Chemical (or Signal), 2, 5, 148

Alkanex wire enamel, 71, 126

American Cyanamid, 13, 23, 124

Anaerobic Permafil, 45-48

Angewandt Chemie, 73

AKU, Dutch company, 87, 94, 166, 183

Antitrust challenge, 67-68

Archer-Daniels-Midland, 44

Arnox epoxide, 105

Asahi Chemical, 111, 148

AT&T, 1

AZDEL, 110-111

Bakelite (Corporation), 9-10, 13, 23

BASF, 3, 5, 109, 141, 147-148, 188, 191,

Bayer, 3, 37, 73-76, 80-81, 92, 94-95, 108, 111, 115-116, 137, 140, 143, 148, 176, 188

"Belt" apparatus for diamond synthesis, 54, 162

Bergen Op Zoom (BOZ), The Netherlands, 94, 98. 105-106, 110, 120, 131, 137, 139, 141, 174, 176, 182-183

Bisphenol A, 70-71, 73, 79, 81, 101, 118, 164

Borazon, 63, 65, 68

Borg Warner, 92, 115-117, 140, 174, 188, 192

Bouncing Putty, (Silly Putty), 18, 21

British Petroleum, BP Chemicals, 100, 109, 145

Business Week, 1996 Conglomerate list, 2

Carboloy Corporation, 20, 43, 186

CD-ROMs, 110-111, 143

Celanese, 5, 6, 92, 96-97, 108

Chemical and Engineering News, 5-6, 107-108, 190-191, 140

Chemical Division formation, 20-26
Chemical industry context, 190
Chemical Materials Department, 21, 45, 76, 91
Chemical vapor deposition process (CVD), 66
Chemical Week, 148
Chevron, 44
Chi-Mei Company, 117
Chlorosilane intermediates, 135-136
Chrysler, 2
Citgo, 110
Commercial Development Centers, 93-94, 95, 110
Compacts, diamond, 65
Compax, 65
Corning Glass Works, 15, 20, 27
Cumene, 100-101
Cycolac resins, 3, 92, 115, 143, 145, 176, 183
Cycoloy alloys, 3, 116-117, 141, 145, 149, 151, 176, 183, 188
De Beers, 4, 56, 60-62, 67-68, 184
Degussa Company, 34, 136
Diamond compacts, 65
Diamond dust, 66
Diamond saw, 56, 62-65, 68
Diamond, synthetic, 44, 49-68, 162, 180
Dimethyldichlorosilane, 18
Diversification, 1-2
Dow Chemical, 3, 5-7, 13, 21, 23, 28, 31, 73, 92, 109, 111, 115-116, 140, 147-148, 190-191
Dow Corning, 4, 20-21, 23, 27-31, 34-35, 37, 40, 129, 131-132, 136, 138
Du Pont, 3, 5-7, 11, 15, 23-24, 35, 92, 108-109, 111, 148, 190-191
Durez, 23
Eastman Chemical, 5, 47, 191
Elchem, 135
Electrical insulation, 10-11
Engineering Plastics, 91-112, 139-158
Enie-Chem Company, 120
Everybody Wins, A Life in Free Enterprise, 190
Exatec Company, 143
Exxon, Exxon Chemical, 1, 5-7, 117, 190-191
Farbenfabriken Bayer (see Bayer)
Financial data summary, 191-192
Flamenol vinyl insulation, 14
Food Machinery Company, 5
Ford, 1, 106
Formex wire enamel, 14-15, 69, 168
Formica Corporation, 23, 123-125
Formvar resin, 14-15
Forty Years of Firsts, 28
Fox melt process, 70-71, 164
GE-Bayer Silicones, 137, 138
GE electromaterials, 3-5, 123-128
GE Plastics-Americas, 145
GE Plastics-Europe, 94, 145
GE Plastics-Japan, Ltd., 118
GE Plastics-Pacific, 145
GE Silicones, 4, 27-44, 129-138
GE organization and culture, 33, 189-190
GE Silicones, 4, 129-138
Geloy, 111, 143, 156, 176
Gem diamond synthesis, 64
Gem Polymers Ltd., 109-110
General Motors, 1
Global Sourcing & Petrochemicals, 145

Glyceryl/glycol terephthalate enamel (Alkanex), 69, 71, 126
Glyptal paint, 11, 128, 186
Goldschmidt, 138
Great Western Silicon Company, 136
Grignard reaction, 16, 28-30
Guaiacol and carbonate, 71
Henkel Company, 48
Hercules Corporation, 5, 100
Hoechst Celanese (Hoechst), 6, 109, 148
Hooker Chemical, 13, 79
Huls, 138
Huntsman, 5, 110, 175, 187, 191
ICI and ICI Americas, 5, 37, 109-111, 191
Iljin Corporation, South Korea, 67
Imperial Chemical Industries, 37
In Search of Excellence, 1
Indian Petrochemicals, Ltd., 140
Insulating Materials Dept. 126-128
International General Electric (IGE) Company, 37, 61, 95, 182
Laminates, 4, 12-13, 21-24, 43, 123-128, 185
Lexan polycarbonate, 69-83, 92, 95-100, 106, 112-113, 128, 140-141, 143, 145, 149, 154, 157, 164, 166, 176, 179, 183-184, 187-188
Living Environments House, 122, 175
Loctite (product), 46-47
Loctite Corporation (American Sealants), 46, 48
Lomod, 111
"Man-Made" diamond, 59-60
M.E. Hogg Autralia Pty. Ltd., 95
Melt process for polycarbonate, 70
Metal-bonded saws, 62

Methylchlorosilanes, 17, 19, 21, 30, 32, 40, 74, 170, 180
Methylsilicone processes, 38
Micamat (mica paper), 126-127
Mitsubishi Engineering Plastics, 111
Mitsubishi Gas Chemicals, 3, 111, 140, 148
Mitsui Petrochemical Company, 79, 118, 165, 175
Mobay, 76-77, 79-81, 108, 111
Monsanto, 5, 6, 13, 23, 76, 81, 92, 108, 115-118
Montedison, 109
M.W. Kellogg Company, 100
Mypolex (diamond dust), 66
National Medal of Technology, 7-8
Nature, 55, 162
Neoprene, 15
New Alchemists, The, 56
Niche strategy, 109
NKK, 136
Norton Company, 57-60
Noryl GTX Blend (PPO/PA), 106-107, 152
Noryl thermoplastic, 83-90, 92, 94-100, 110, 111-113, 143, 145, 149, 152, 155, 164, 166, 179-180, 182, 183, 187-188
Nylon, 15
Organization evolution, 145-146
Owens Corning Corporation, 139
Permafil varnishes, 45, 48, 127
Permatex, 47
Personnel practices, 63, 77, 87
Phenol plant, Mt. Vernon, 100-101, 110
Phenol plant, Pittsfield, 13, 21, 24
Phenolic resins, 11-12, 21, 24, 43, 84, 109

Phosgene 73, 79, 81, 105-106, 118, 120, 176

Plastics applications, 149-158

Plastics Laboratory, 11-12

Plastics Department, 21, 91-92

Polycarbonate development, 71-73

Polycarbonate processes, 70-71, 73, 79-80, 100, 118, 120, 140

Polychlorinated biphenyls, 136

Polymerland, 3, 117, 140, 145, 176, 188

Polymer Processing Center, 120, 122

Polyphenylene oxide (PPO), 84-90, 92-100, 109-111, 164, 166, 172, 174, 180

Praxair, 5, 191

Pressure-sensitive adhesives, 135

Process of Innovation in the Chemical Industry, The, 190

Product line evolution, 141-145

QQC Company, 66

Research Laboratory (and CR&D), 9-10, 13-21, 25-26, 43-45, 53-56, 69-71, 83-84, 98, 103-105, 112, 146, 168-169, 189

Resinmec, 141

Resins and Insulating Materials (RIM) 11, 15, 18-19, 20-24, 32, 126

Rhodia, (also Rhone Poulenc) 4, 37, 40, 137-138

Rochow direct process, 23, 38

Room-temperature vulcanizing (RTV) silicone, 40-41, 129-131

Sealants, RTV rubber, 129-131

Shawinigan Resins Company, 14

Shell Chemical, 73, 109, 145

Shin Etsu, 4, 37, 131, 138

Silicone rubber, 19-21, 28, 32-34, 38

Silicones, 34

Silicones and their Uses, 28

Silicone resins, specialties, 135

Silicone project, 15-21

Silly Putty, 18, 21

Solvay, 109

Specialty Chemicals, 145

Specialty Materials Department, 62-63

Stauffer Chemical, 41-42, 111, 132

"Stick to the Knitting", 1, 8

Stratapax, diamond drill compacts, 65

Sulfonated polystyrene, 12

Supec, 111

Superabrasives, 4

Synthesis of diamond, 52-56

Teijin Chemicals, 111, 140, 147-148

Terephthalic acid synthesis, 83

Textolite, 13, 124-125, 185

Thermodynamics and high-pressure science, 50-52

Thermoplastics, 44, 72, 93, 107-109

TIPCO, 140

Toray, 148

Toray-Dow Corning, 138

Toshiba, 37

Toshiba-GE Silicones, Tosil, 131, 138

Ultem polyetherimide plastic, 103-105, 143, 145, 166, 174, 180

Union Carbide, 5, 13, 23-24, 37, 39, 73, 109, 134, 137, 191

U.S. Rubber Company, 29, 92

Valox polyester plastics 96-101, 105, 112-113, 143, 145, 153, 164, 166, 172, 176, 180, 184, 187

Vinyl chloride polymers, 14
Vinyl chloride polymers, 14
Von Roll Isola USA Inc., 128
Wacker Chemie, 4, 132, 137-138
Waste disposal, 136
World silicone industry, 137-138
Wire enamels (Formex), 14-15, 69, 168
Witco, 4, 134, 137-138

Woodhill Chemical, 47
Works laboratories, 9, 10, 12, 14, 69, 71, 88, 189
W.R. Grace, 5
Xenoy blend, 106, 113, 143, 184
Zytel, Du Pont polyamide, 72, 92, 108